THE TRIUMPH OF GOLD

Also by Dr. Franz Pick:

The US Dollar: An Advance Obituary
Third Edition (1986) $70.00

All the Monies of the World (1971)
      $80.00

For further information or to order, please write P.O. Box 40, Bethel, CT 06801

Copyright 1986, by the The Institute for the Preservation of Wealth, Inc., P.O. Box 40, Bethel, CT 06801. All rights to this book are reserved. No part of this book may be used or reproduced in any manner whatsoever without written permission. ISBN: 0-938689-01-0

Second printing: 6/1986	25,000
Third printing: 7/1986	22,000
Fourth printing: 8/1986	25,000
Fifth printing: 9/1986	5,000
Sixth printing: 10/1986	5,000
Seventh printing: 1/1987	5,000
Eighth printing: 4/1987	5,000
Ninth printing: 6/1987	5,000
Tenth printing: 9/1987	5,000
Eleventh printing: 8/1989	5,000

THE TRIUMPH OF GOLD

By Dr. Franz Pick

Introduction by Dr. Henry G. Jarecki

The Institute for the Preservation of Wealth, Inc.

THE
TRIUMPH
OF
GOLD

By Dr. Franz Pick

Introduction by Dr. Henry G. Jarecki

The Institute for the Preservation of Wealth, Inc.

Table of contents

Introduction 1
    by Dr. Henry G. Jarecki

1. Why the head of the Federal Reserve requires the services of a bodyguard 14

2. Why the government's economic statistics are nonsense . 15

3. Three steps to inflation 17

4. Volcker has not eradicated inflation — he has simply bought a little time 19

5. The politicians in charge of our currency live in a make-believe world 20

6. America should declare bankruptcy 21

7. A small secret group with ''emergency'' powers will decree the currency exchange 22

8. We will exchange at least 100 old dollars for one new dollar 23

9. Social Security will go down the tubes 24

10. Collectibles are not a reliable hedge against inflation 25

11. Currency theory opens up the dark drawers of official corruption 26

12. The 3 fraudulent US government bankruptcies 27

13. The orchestrated press embellishes the value of the dollar 28

14. Gold has performed better than most holdings 29

15. There is simply not enough money to pay for all the goodies that have been legislated 30

16. The deterioration of the dollar is destroying America's political superiority 31

17. The gross national product has not risen in 45 years 32

18. The Russians have more gold than we do 34

19. The dollar was wiped out once before 35

20. The US government will crawl into every safe-deposit box to get at your gold 36

21.	You will be lucky if you can take $500 out of the country	38
22.	Government loses battle in anti-gold war	39
23.	The $12 trillion total indebtedness in the US will never be repaid	40
24.	The dollar is doomed	41
25.	Stamps are not a hedge against inflation	43
26.	Inflation has eaten away the golden marrow of the dollar	44
27.	Real estate is a disaster area	45
28.	A trick for preserving your estate intact for your heirs	46
29.	Gold in Switzerland is safe	47
30.	The effect of a currency exchange on dollar-denominated assets	48
31.	The most practical gold coins for the coming currency crisis	49
32.	Governments and corporations prefer to hide the truth	50
33.	How to buy gold in London	51
34.	There will be two currency exchanges: the first will be benign, the second malevolent	52
35.	American currency policy since 1940 has been nothing but an unbroken series of flops	54
36.	The only gold holding that can be kept in your safe-deposit box	56
37.	Our assets are melting away before our very eyes	57
38.	Few people understand the concept of currency debasement	58
39.	International transfers of currency will require a special license	59
40.	Our ''conservative'' president will not be able to stop inflation	61
41.	Bah. We will not have deflation in the United States	63
42.	Sooner or later all monies die	66
43.	The danger in opening a safe-deposit box under an alias	68

44. Since 1973, the value of worldwide gold hoards has risen 301% 69
45. Fully paid holdings of gold bars and coins should not be sold 70
46. The stock market is a graveyard — but they refuse to bury the bodies 71
47. The history of currencies is not sympathetic to men who tried to stabilize the money with deflation 73
48. I recommend krugerrands to protect you from my other clients 75
49. If all the silver in Washington's stockpile is sold, the profits will not wash away a speck of red ink from the budget 76
50. How a paper money panic affects investment demand for gold 77
51. The United States came close to a total financial collapse in early November 1978 78
52. The new paper currency will be tied to gold 80
53. If the true condition of the US corporation were known, no one would buy stocks or bonds 81
54. A loaf of bread will cost $100 83
55. Since 1933, the US budget has been an instrument of electoral bribery 84
56. How to buy gold in Switzerland 85
57. Inflation under 10% is merely an illusion 86
58. 40 years of dollar debasement cannot suddenly be reversed 87
59. Get some of your gold out of the country NOW — before it's too late to do it legally 89
60. Two metals which are underpriced 91
61. Silver will not be confiscated when they come for your gold 92
62. When the currency exchange comes, the suffering will be terrible 93

63. The minidollar is ripe for a ruthless liquidation 94
64. People believe they can live with inflation forever — but one day the bubble will burst 96
65. Inflation has made ''investment'' a pornographic word . 98
66. There are only two sensible uses for diamonds 99
67. I will never go short gold, nor will I ever be out of the gold market 100
68. The ''Great American Inflation'' is the result of willful abuse of the creation of money and credit 101
69. Government bonds are certificates of guaranteed confiscation 102
70. If the riots in South Africa continue, I am afraid the gold production will severely decline 103
71. The currency problem, as I see it, is a rape of the law 104
72. My tailor knows more about currencies than all the idiots at the US Treasury put together 105
73. By debasing the currency, we have endangered the economic existence of the United States 106
74. Corporate profits, on a per capita basis, are only about a quarter of what they were in 1950 107
75. I itch in my hand to buy gold now 108
76. The biggest single industry in the United States is corruption 109
77. I do not give any country the right to limit the transferability of my currency 110
78. The top 3 or 4 banks will be officially nationalized 111
79. The Eurodollar has to be compared to the second use of toilet paper 113
80. We will have a new ''hard'' dollar on a gold standard 114
81. Why corporations should use constant dollar balance sheets 115
82. How currencies die 116

83. The pious pronouncements to hold the money supply in check will not be kept 117

84. Bond salesmen's propaganda that ''a dollar is a dollar'' should be rewritten to say ''a dollar is 3¢'' 118

85. The US government should build a monument to Johannes Gutenberg 119

86. The triumph of gold 120

Introduction
by Dr. Henry G. Jarecki

Franz Pick, the world's foremost gold bug, is dead. A prolific writer and theoretician for the last sixty-five years, he was for thirty years the publisher of the <u>Black Market Yearbook</u> which, in deference to the respectability of his central banker clients, he renamed <u>Pick's Currency Yearbook.</u> He and I were friends from the time he directed me in 1967 to the bullion firm I now lead until his death last week. What I admired about him was not his predictive ability but his irascibility, energy, and rigid morality. True enough, he predicted (or said he did) the 1926 devaluation of the French franc, the 1931 devaluation of the British pound, and the meteoric rise in gold that emerged in the late '60s. But those predictions were not, when one knew him, predictions at all. They were rather the words of Hosea declaring that the children of Sodom and Gomorrah were behaving badly and would thus come to no good. They were not, any more than Hosea's words were, practical guideposts. Not at all. He knew no more about when gold or currencies would go up or down than anyone else can in this random world.

He and I often argued about its randomness, I saying all of life was unpredictable; he saying it was obvious destiny. To defeat my scepticism, he repeatedly cited his two somewhat diverse educational inputs: his hometown rabbi whose Talmudic teachings proved, for Pick, that every currency was doomed to be unstable; and, second, the nineteenth-century Prussian economist Georg Friedrich Knapp, who believed that money was an instrument of state order, that its value reflected the public's attitude toward the state, and that — given man's failings and thus the state's — the only road a currency can go is down.

Pick told me:

The most important prediction I made was very romantic. In August 1931, my new wife Peggy and I lived in Paris. That summer, she went to Bohemia to visit my parents and, knowing she would be gone for a month, I went sailing. I sailed from Houlgate (the small French village where William of Normandy first gathered with the fifteen men who went with him to Hastings in 1066) to England, but when I arrived there — in the small town of Cowes, to be exact — I found so many other boats there that I could not moor. Small boats, one after the other, were bringing sailors to shore from nearby battleships.

I asked one of the sailors, ''Why?''

He said, ''Young man, we're going on strike.''

''Why?'' I asked.

''For higher pay,'' he said.

Knapp had taught me that currencies fall when there is a rupture in the state and I thought striking sailors were a manifestation of such disorder. I concluded that the currency had to fall and so I ran to the telephone and called the Czech financier, Petschek, to tell him to sell the pound sterling short. That was on August 10, 1931. Petschek sold one million pounds short — half to the Czech National Bank and half to Kleinwort's in London — on Yom Kippur (which happened that year at the equinox). On September 21, 1931, sterling was devalued and Petschek made a million dollars.

''Why do you mention the equinox?'' I asked

this hyperrationalist.

''I have a theory about devaluations on the equinox. I published it once in a book.

''I learned currency from Knapp,'' Pick continued, referring to Georg Friedrich Knapp, the Prussian professor of monetary theory who wrote that *Das Geld is ein Geschopf der Rechtsordnung* (money is a creation of juridical order).

Pick and I enjoyed arguing about Knapp. I found Knapp's work esoteric and hard to understand and I disliked his use of such incomprehensible terms as ''accessory'' and ''valutory'' units of money. But that was not my main objection to Knapp. I also held him partly responsible for the 1923 German inflation.

He had been the teacher of Helfferich, the Finance Minister who presided over that period and shared Knapp's view that money was worth whatever the state said it was. Even worse, he thought that Germany's prices were so high because there wasn't enough money in circulation and that prices would come down if only he printed more. Such ideas seem extremely dangerous to people who think, as I do, that ''money is a feeling in the stomachs of the people,'' and not what the state says it is; for if the state keeps printing, the people will lose confidence and the money may not be worth anything at all.

And yet, despite our disputes, I found Pick's Prusso-rabbinical currency views sound and stimulating. I knew him in his later years, when he was plagued by arthritis, stomach cancer, cataracts, and, yes, some rigidity in his thinking and aphorisms. But the host of proverbs and aphorisms with which he could brighten a Sunday afternoon marched in a straight and internally consistent file as if they were soldiers in Emperor Franz Joseph's Royal Imperial Army, the only army and monarch Pick ever served.

You could hear the rabbi, the Prussian professor, and even the Emperor himself when Pick thundered out: ''As long as a judgeship sells in New York for $350,000, the dollar cannot continue to exist.'' One of the longest-standing disputes between Pick and myself was whether his views were moral or practical. I said that his aphorisms were so overwhelmingly moral in content that they were almost useless as guideposts. I told him I thought of him as Hosea claiming that the children of Israel would come to no good. I added that a prophet should not simultaneously thunder warnings and define how things should go. He disagreed.

Pick was so sure of man's weakness and of the consequences of this weakness on the state and its currency that he was sure the only way a currency ever does go — or can go — is to depreciation. ''At its creation,'' he often said, ''the only destiny a currency has is devaluation.''

He often repeated himself, especially in his later years. And yet he did not insist on the truths in which he believed. He was satisfied that they would out by themselves. Time and again, he told of his visit to Carl Jung's home in Küsnacht where the inscription over the gate says, *Vocatus atque non vocatus, Deus adherit* (Whether you call him or do not call him, God will be there). Again and again, he used this phrase to say that economic facts, whether identified or not, will prevail.

Pick was born in 1898 in Boehmisch Leipa on the Judengraben. His mother died young — on his Bar Mitzvah, to be exact — but his father, a traveling salesman, died at the age of 86. His only brother died in Auschwitz in 1944. Pick left Boehmisch Leipa for the Army when he was 18, then moved to Leipzig and Hamburg and later to France.

Pick served in the Austrian army until 1918, when he was wounded on the Italian front. After the war, he went home to use the fund his father had set aside at his birth to pay for his university studies. Unfortunately, it could, by the time he needed it, buy only two meals and not the full education for which it had been planned. And so when he went to the University of Leipzig to study economics in 1919, he had to moonlight as a copy reader for a Leipzig newspaper. In 1922, he transferred to the University of Hamburg to study under Dr. Kurt Singer, a student of Knapp, who introduced him to the master's works. Pick's dissertation was classically Knappian: ''Austrian Monetary Policy During the Napoleonic Wars: a contribution to the governmental theory of money.'' In Hamburg, too, he earned his keep by working nights at a local newspaper.

In 1923, he received a Doctorate in Political Science and moved to Paris to work for Louis Dreyfus and Company for 500 French francs a month (then worth about $20). His work went well for the first two and one-half years, but he was ultimately betrayed by his long-standing habit of newspaper moonlighting (which in Paris he used to support the sailing habit he had developed at age six in Boehmisch Leipa by setting sail in his mother's washtub on the town's Deer Park Pool). While working for Dreyfus, he wrote a syndicated column called ''Die Weltboersen'' (The World's Stock Markets) which predicted (correctly as it turned out) the devaluation of the French franc. Dreyfus, concerned that this reflected on them, promptly fired him.

Pick never took on another job. Instead, he worked as a self-employed currency advisor, journalist, and author whose first book, <u>Sense and Nonsense of the Stock Exchange,</u> was published by S. Fischer of Berlin in 1929.

In 1931, he married Peggy Robson, a tourist from Chicago to whom he had predicted their marriage two hours after they met. They emigrated to the United States in 1940, first to Chicago and then to New York where he remained ever after.

From the very beginning, his roles as currency advisor and journalist overlapped: Colonel McCormick's <u>Chicago Tribune</u> owned a company called Press Wireless in Paris. When its crew was left penniless in the throes of a French currency crisis, McCormick asked the Foreign Exchange Manager of the First National Bank in Chicago what to do, and he in turn asked Pick. Pick contacted Silverman, the owner of a fashionable fur shop on the Rue Edward VII, and asked him if he had any money in France. ''Sadly enough,'' said Silverman, ''I have a few hundred thousand dollars there.'' Happily enough for Pick, however, for he got Silverman to pay the Press Wireless staff and McCormick to pay Pick $300 a month for the year or two the arrangement lasted. Soon after, another advisory account came his way and they kept coming ever after.

Pick's first American article appeared in <u>BARRON'S</u> in December 1940 under the prescient name ''If a Blackout Came to the United States, What Would It Do to the Financial and Stock Markets?'' Soon after the war ended, <u>BARRON'S</u> asked him to write other articles on the black markets then springing up overseas. These gradually evolved into the <u>Black Market Yearbook,</u> later called <u>Pick's Currency Yearbook.</u>

In 1961, he started to give seminars. His first was a ten week currency seminar he held for twenty-two junior executives at a restaurant in downtown New York. Since then, he has given hundreds of such seminars, usually for sixty to seventy people at a time. His style is dramatic, bombastic, shocking, charismatic, and prophetic. His listeners love or hate him — and even those who hate what he says love the aphoristic, spellbinding way he says it.

Once a year, he spoke to 4,000 or more followers at the National Committee for Monetary Reform in New Orleans. His followers are legion: small businessmen, doctors, lawyers, and the like. They loved to ask him about ''investments'' and loved it when he berated them as stupid or shouted, ''There is no such thing as an investment.'' He would show them the mortality table of currencies to demonstrate that no currency ever lasted long enough to make the yield one gets on it worthwhile. When his clients asked him if they should buy bonds he would tell them, ''Every government bond is a certificate of guaranteed confiscation'', and remind them of the words of the Mishna Baba Meziah: *Thou shalt not buy a debt for more than one tenth of its value.* ''Based on that,'' he winked at me, ''I bought a number of defaulted government debts and made money.''

He would emphasize his scathing views about bonds by recalling the time Paul Reynaud, former Prime Minister of France but then out of office, invited him to the Chamber of Deputies to hear Reynaud attack Leon Blum, then Prime Minister of France, in words Pick has repeated ever since: *To the owners of government bonds, you have left only the eyes to cry.* (Sometimes he cited it: *You have not left them the handkerchiefs with which to dry their eyes.*)

He was not generally well-liked by central or

commercial bankers. Once when I introduced him to the Number Two man of a prestigious Middle European bank who was spending the weekend with me, he shouted at the startled man, ''Your chairman is a swine.''

Reflecting on his style and message, Dr. Pick commented:

They are, in part, the result of the fact that a few of my teachers were Jesuits. And, until the day he died, Pablo Arupe — the man who was called the Black Pope — the general of the Jesuits, was my client.

It is no wonder you think I am a moralist, Dr. Jarecki. I am the product of four men: the rabbi of my youth; Georg Friedrich Knapp; Lord Keynes; and Charles Rist, who wrote <u>The History of Economic Opinions.</u> Their teachings have made me what I am today: the only man in the country who understands monetary theory.

Roosevelt prohibited the teachings of monetary theory in this country, and today there is no chair for monetary theory anywhere, not even in Switzerland. Why? Because no government wants its people to hear the truth.

In November 1939, Radio Berlin said I had been sentenced to death for my writings. From 1952 to 1955, I was on the Bank of England's Blacklist for publishing a list called ''The 89 Varieties of Sterling.''

But all was forgiven, I think, for the British Chancellor of the Exchequer and I attended a seminar together in Southern England some years later and he said to me, ''Pick, you were wrong. You forgot one I created, namely, petroleum sterling.'' And in the United States, I have been called before the federal grand jury for writing a book that showed Americans how they could safeguard their money by having Swiss bank accounts.

The authorities had good reason to find his comments offensive. In 1950, when he first published the <u>Black Market Yearbook,</u> he dedicated it to ''the more than two billion victims of inflation who, for obeying the law, have been punished by the law,'' a phrase he took from the words written on the Columns of Thermopylae. In a later version, he dedicated it ''to the central banks of the world, without whose cooperation the black markets described in this book would not have been possible.'' In time, however, the annual summary had become so useful that it had to become respectable. In 1952, at the request of clients who were embarrassed to have such an oddly-named book on their shelves, he renamed it <u>Pick's Currency Yearbook,</u> under which title it was published until 1982. ''I have written 71 books; I used to publish the list in my yearbook. But the topic I was proudest of is the one that more and more public corporations have now adopted: how to maintain balance sheets in constant dollars.''

My own introduction to Dr. Pick came through the book called <u>Silver, How and Where to Buy and Hold It.</u> I bought this book when I saw it advertised in the <u>NEW YORK TIMES</u> for ''thirty-five paper dollars a copy.'' I found it to be a very practical book on exactly how one can transact in the silver market. It included sections on how one could buy and sell silver — what the invoice and transactions sheets would look like when the mythical ABC Bullion Company of 123 Park Avenue bought silver from the equally fictitious XYZ Silver Company of 345 Madison Avenue.

One chapter was called ''London Silver'', and in it I saw another one of those obviously phony invoices replete with letterhead and all from a company I knew could not exist: Mocatta & Goldsmid. ''Imagine that,'' I thought, ''calling a silver company Goldsmid.'' At the same time, however, I had just gotten an international direct distance dialing telephone. When I had a particular issue that I wanted to discuss with a bullion dealer in England, I kept playing with that phone and finally decided to spend the $9.00 for three minutes, quite sure that if I called the number that was on that letterhead I would only find a little girl answering and saying that her mother was not in. To my surprise, I found on the other end of that phone Keith Smith, then a dealer at and today the Managing Director of Mocatta & Goldsmid. Smith and I quickly got into a conversation that led to the development of a substantial silver business and ultimately to my association with the Mocatta Group.

But Pick was disinterested in business as such:

I was never in business except that I was a paymaster of the European Resistance. And I don't even like to talk about that. As Goethe said: *Politisches Lied, Welch Garstiges Lied* (the political song is a horrible song). That was, in any case, my second military affair. When I was 19 and a first lieutenant in the Austrian army, my commandant claimed I was having an affair with his wife and challenged me to a duel.

Pick, whose romantic life — even after his wife Peggy's death in 1974 — remained a central point in his adventurous life, smiles at a memory:

I still have a bullet fragment in my ear as a souvenir of that episode. One other effect of the duel was my symbol. You see, in those days the proper uniform for a duel was, of course, black pants and a white shirt. I embroidered a black Ace of Spades (in German, the Ace of Spades is called *Pik As*) on my shirt and it has been my trademark — and the name of every one of my boats — ever since.

Writing, adventure, and public morality were his strengths. And so it was no wonder that Vietnam raised his ire as a cause of the depreciation of the American currency:

> Vietnam cost this country 600 billion minidollars, of which 300 billion were stolen outright, primarily by sergeant and generals.

Again on public morality: ''There are only two countries where judges cannot be bought — Switzerland and Great Britain.'' Among the most corrupt, he often said, were Israel and the United States. ''I am a devout Jew,'' he said. ''I can conjugate the irregular verbs in Hebrew; the only thing I didn't get in my life was the Kabbalah, but I believe that there is great corruption in Israel.''

And again:

> About 15 years ago, I wrote to William McChesney Martin, Chairman of the Federal Reserve Board, and told him I wanted to see him. Two days later, he set up an appointment. When I entered his office, I said to him, ''Before I sit down, I want to make a statement: I have no favor to ask of you and I have nothing to sell.''
>
> ''That has never happened to me before,'' said Martin.
>
> ''I have come to help.''
>
> ''Go ahead and talk.''

I told him, ''I suggest that you float a $25 billion bond issue at 2½% interest for twenty-five years without repayment. No questions asked, free of income or inheritance tax; principal and coupons linked to the cost of living.''

Martin got up, paced back and forth, and said, ''Franz, it is politically impossible.''

We have remained friends ever since. I admitted to him that the idea was not my own, but that it came from an old friend of mine who ran a cannery in southeastern France. That old friend is Antoine Pinay who, as Finance Minister of France, issued such bonds to the grateful people of France. Pinay bonds exist to this day and have doubled in value since they were issued, something that cannot be said about any other bonds. ''If you do this,'' I said to Martin, ''your main buyers will be the Russians and Chinese. If you don't, you will capsize your currency.''

I summarized for him a few years before he died:

If I understand you correctly, Dr. Pick, currencies depreciate because of corruption: the British sailors, the corrupt judges, the corrupt American sergeants and generals in Vietnam, the President who was unwilling to get Congressional and popular support for the money he wanted to spend in Vietnam, and the people of every country who want roads, schools, bread, and circuses but don't want to pay the taxes with which to get them.

He agreed, but added that it was not the mini-corruption of sergeants, generals, sailors, and judges alone which demonstrated the falling apart of the community; but also the more legalized corruption in which elected officials grant favors to everyone. The Congress, so Pick said, gives something to everyone who comes with a plea or a contribution. And so, Pick said, his estimate that the cost of corruption in the United States is trillions of dollars a year is vastly understated for it does not take into account such elements of corruption as minimum wage laws, government support of unions and industries, bailouts of Chrysler and Continental and all the rest.

And, he said, one must add to all this corruption the confusion that the modern financial system permits — for this, too, is needed to let it all happen:

> A multiplication of the instruments of payments is one of the ways the depreciation occurs. I speak here particularly of those nonsense instruments called SDR's and of the enormous development of the Eurodollars.

I asked him whether a return to the gold standard and/or to a 100-to-1 devaluation would help. He said he thought it was possible — that it would be very bullish for the economy because the entire corporate debt would be wiped out, and because the 32 million civil servants would rapidly have the new currency and so give rise to a great upsurge in the economy.

Pick was fully in this age, but he was unmistakably of the old school: urbane, cultured, articulate, cynical, righteous, and square. He was a friend of mine and of our time. His words, alas, like Hosea's, will be remembered for centuries to come.

Dr. Henry G. Jarecki
Chairman, Mocatta Metals Corporation
Chief Executive, The Mocatta Group

December 5, 1985

1. Why the head of the Federal Reserve requires the services of a bodyguard

Whether or not he is aware of the incidents affecting the lives and fortunes of his peers in currency history, it is a well-known fact that the current head of the Federal Reserve System does not appear in public or travel anywhere without the escort services of a bodyguard.

As for the present Secretary of the Treasury, his permanent entourage includes a platoon of agents assigned to protect his life and limbs.

Life in twentieth-century America can be as violent as it was during the French revolution or in post-World War I Czechoslovakia and Germany. The tortures being inflicted in the name of financial and currency morality may well be compared to the medieval acts of moral purification. It remains to be seen for how long ''grateful'' Americans will tolerate economic and monetary mortification.

2. Why the government's economic statistics are nonsense

The subterranean economy — the undocumented, untaxed, illegal, corrupt portion of the economy — accounts, by my estimate, for between $900 billion and $1 trillion: 25% of the gross national product. This is one of the keys to the destruction of the dollar.

The largest sector of this subterranean economy is probably the hard narcotics. My colleagues and I estimate that the total hard narcotics trade is from $300 to $500 billion. We cannot say exactly how much it is, because it doesn't go by check. It goes by diplomatic valise to Switzerland and the Caribbean banking centers.

There are four centers of hard narcotics trade: New York, Los Angeles, Chicago, and Miami. The little black kid in Harlem pays three dollars for the fix that he sells for $35. The fix is half heroin and half milk-sugar: very low grade. The people who put it into the glassine envelope make much more money.

What we get here in New York comes through Miami from Bolivia, Costa Rica, and a few other countries. People there pile up billions of dollars in illicit profits.

The people who make the billions in graft and corruption do not own government bonds. They do not pay taxes. Do you see the significance?

The whole subterranean part of the economy is not included in the official statistics. So they give the wrong picture of the economy. It means the government doesn't have the pertinent statistics to judge the economic problems of the country. It means that all the government's economic theories are nonsense.

If we could tax this sector, we would not have a budget deficit. We would have a surplus. The taxes that are avoided are cash that has to be printed. The Treasury must float bonds, officially, in order to cover it and create the necessary income.

3. Three steps to inflation

Before Roosevelt, in the 1930s, a person made it or failed on his own. If he failed, it wasn't the government's fault, or society's. The person was responsible for himself.

But with Roosevelt, if you failed, it wasn't your fault — it was the government's. So the government would take care of you. This is where inflation starts.

The next step is when the government has to separate the currency from gold. That was Nixon, in 1971, when he closed the gold window. So you see, the destruction of the dollar is bipartisan.

Once the dollar was separated from gold, there was no restraint on the government. That was the green light for the Treasury to debase the dollar at will.

As the inflation deepens, people try to protect their assets. They take their money out of the country. So the third step is exchange controls. It starts slowly, then bit by bit, the government turns the screws tighter.

This third step went into effect in 1972. That was Public Law 91-508, which requires you to report to customs when you come into or leave the US with more than $5,000 cash or other monetary instruments. Then, the banks have to report your deposits or withdrawals of $10,000 or more, in cash, to the Treasury.

They even have the power now to open up first class mail going overseas — if they think you're trying to avoid the currency restrictions. Not many Americans know that. But they know you have to report a foreign bank account on your income tax return. So, you see, little by little, the exchange controls tighten.

4. Volcker has not eradicated inflation — he has simply bought a little time

Only fools believe the current remission from inflation is permanent.

When I was a child in Bohemia, my father took an endowment policy that would see me through four years in the University. Just 15 years later, the insurance company paid me off in full with paper currency that barely covered the cost of my first two meals in the University.

So I speak from personal knowledge. I lived through the destruction of the Czech koruna, the German mark, and the French franc. Always there were periods of time in which inflation slowed, or stopped. Always inflation came back worse than ever. I see that happening here.

So with this slow-down, Volcker has not eradicated inflation. He has simply bought a little more time. The dollar will be wiped out.

We have had constant, creeping inflation for the dollar since 1940. Because the dollar was the kingpin of the world's monetary system, like venereal disease, it infected all the other currencies of the world. Soon, we too will go through the wringer. I only hope I am no longer here. I am too old and tired for such a crisis.

5. The politicians in charge of our currency live in a make-believe world

We are living in a period of horrible confiscation of people's assets through inflation. The dollar is being destroyed.

The dollar is now worth between 2 and 4 pennies of its 1940 value. The regime in Washington is trying to save the dollar, yet there is not one person in the US Treasury who knows anything about currency. It's the same as if you broke a leg and went to a doctor and he said, ''I know nothing about anatomy, but I will set it.''

It is no accident that the same familiar faces of money moguls can always be noticed at the fringes of the Oval Office, regardless of who the occupant may be or his publicly professed politics.

Therefore, all attempts by the regime in Washington to cope with the debasement of the minidollar will, in the end, prove to be in vain. Whatever successes it claims are short-lived. And they have been achieved more by accident than by design.

6. America should declare bankruptcy

Numerous signs of monetary and political decay confront the ruler of the land. Mastering inflation and unemployment — along with a return to increasing prosperity — is the official *leitmotif* of the current occupants of the top level of government. But this has been the theme song of American politics since the Great Depression of five decades ago — which has haunted and corrupted the thinking of America's intelligentsia.

''No depression at any price'' has become the holy writ of US political economics since 1945. And it has made the inflationary way of life the *ne plus ultra* of economic and financial activity in this country.

Should the ruler avoid biting into the always bitter apple of state bankruptcy, the refusal to stop monetary and price inflation will bring the country to higher levels of inflation — with the debasement of the currency approaching near-panic levels.

Furthermore, the continued absence of monetary stability would cause oil-producing countries to increase the price of petroleum automatically with every new increase in the cost of living in the United States.

Not only oil and precious metals, but also all other commodities would undergo rapid price increases.

Therefore, there would be a snowballing effect on all prices, including wages and service fees, paralleling the automatic decrease in the standard of living for all those dependent upon pensions or annuities — along with the breakdown of all institutions living off endowments.

7. A small secret group with "emergency" powers will decree the currency exchange

The freezing of Iranian assets in late '79 and the closing of grain markets in early 1980 are a foretaste of things to come.

When the currency exchange takes place, foreign countries will stop accepting US dollars. Some of the OPEC lands could ask for a gold guarantee for the dollar assets in Western banks — or perhaps could just stop accepting dollars for its goods and services.

American assets abroad could easily become subject to retaliation. Visible American property of any kind in foreign countries could be blocked for the time being. This, in turn, will add to the difficulties of settling the dollar problem.

So you see, above all, action has to be swift. Therefore, Congress could not legislate it. It has to be a small secret group with full ''emergency'' powers to decree new and unpleasant monetary rules. This probably includes a ''temporary'' moratorium on all debts.

8. We will exchange at least 100 old dollars for one new dollar

A storm is gathering around the dollar. We live in a make-believe world. One day it will have to crash.

We will have a new currency. We will exchange at least 100 old dollars for one new dollar. Maybe 1,000. I do not know which.

I have told you the dollar is at approximately 3 pennies of its 1940 value. Call me a liar if it's 2½ or 3½. I establish my figures an average of once a week at the supermarket.

We cannot go lower than about 2 pennies. Because then the automobiles would cost $30,000 a car. It would be burdensome.

In a 100-to-1 currency exchange, one unit of the new currency will have the same purchasing power as 100 of the old units. All we are doing is making bookkeeping changes, knocking off zeros, and officially acknowledging the rape of the currency.

The day of currency exchange signifies that the government embezzlers have been successful in expropriating almost all of the value of the dollar. The only one who will laugh on the way to the bank will be the US Treasury.

9. Social Security will go down the tubes

When the currency exchange takes place, Social Security and pensions will go down the tubes. On occasions in the past, such as during the European currency changeovers, they readjusted the private and public pensions. But don't count on it.

It's a political horserace. How can you predict the actions of, let's say, the not exactly clean congresspeople?

Social Security is political. It may be adjusted upwards. But then we'll have the same currency problems again.

10. Collectibles are not a reliable hedge against inflation

I have a Jackson Pollack painting, which he gave to me, dedicated to Peggy and Franz. It didn't cost me a penny. A museum offered me 80,000 minidollars for it.

But there are disappointments too. I have a painting by Brueghel — ''The Adoration of the Magi'' — which I bought nearly 60 years ago for $5,500. In 1978, I could have sold it for $800,000. Today, I couldn't get $250,000 for it.

So, collectibles are not a reliable hedge against inflation.

11. Currency theory opens up the dark drawers of official corruption

The underworld is so strong that possibly there is not one congressman or senator who is clean. These people are not elected with their own money. Somebody paid for their elections and they are the servants of the people who paid for it.

Currency theory opens up the dark drawers of the official corruption and bribery. I lived through, as well as witnessed, both of Germany's runaway inflations: after World War I and after World War II. Even in Germany, you could at least teach currency theory.

But here, you can't. The teaching of currency theory in the United States fell into disuse under Roosevelt. Today, no college, not even Harvard, would allow a course in currency theory. Because if currency theory were taught, the issuance of government bonds would be prohibited — as it should be.

12. The 3 fraudulent US government bankruptcies

There have been 3 fraudulent state bankruptcies in the history of the United States. The first was in January 1934. The official price of gold jumped from $20.67 per ounce to $35.

The second dollar devaluation, a comic book event, took place in December 1971. The official price of gold increased from $35 per ounce to $38.

Then, on February 12, 1973, the dollar was devalued for the third time. The official price of gold increased to $42.22 per ounce.

The two most recent devaluations alone amount to about 18%. If we continue to do this, we are going to ruin the United States — and we may drift into dictatorship.

The destiny of the currency is, and always will be, the destiny of the nation.

13. The orchestrated press embellishes the value of the dollar

I cannot tell when the currency reforms will come. It may be one year, or as long as ten years.

We hesitate to accept the currency exchange. We try to postpone it with all kinds of public relations gimmicks. The whole orchestrated establishment press in the United States embellishes the value of the dollar. If the gold price goes down $1 an ounce, the New York Times uses 20 or 30 point type. If gold goes up $5 an ounce, they use 6 point type.

But the currency exchange will come. It must — because no ruler or government in history has ever mastered, or wanted to master, the principles of a stable currency, and neither will the present government.

14. Gold has performed better than most holdings

We are probably living with more or less stable gold prices for the time being. But it would not surprise me that as soon as the ruler's budget and tax cuts are approved and fervently acclaimed by one and all in Washington, the gold price will begin climbing — a worrisome omen of things to come. We will eventually see $1,000 gold.

I love my assets in gold and will not sell them. Gold has performed better than most holdings.

15. There is simply not enough money to pay for all the goodies that have been legislated

The trillion minidollar debt behemoth has appeared with nary a whimper being heard. Like a Biblical plague, it has been supinely accepted by one and all as the will of God.

This negative achievement of the hopelessly mismanaged finances of the United States government merited only fleeting mention in the orchestrated press.

All the analyses presented for public consumption could be boiled down to the soothing platitude that, after all, ''we owe it to ourselves.'' This line has been trotted out for over 40 years to justify the deficit financing and inflationary process.

But the truth is, there is simply not enough money coming in to pay for the goodies that have been legislated for more than 4 decades — all in the name of prosperity in perpetuity. The deficits in the budget are currently being covered by an outpouring of short-term debt certificates. The soaring interest rates that Washington has willingly paid for such money represent a ''bankruptcy premium.''

16. The deterioration of the dollar is destroying America's political superiority

The current administration floated into Washington five years ago on a tide of hopes that the monetary crimes which had been committed against the dollar for the past 40 years would soon be undone.

New buzzwords were invented and magical incantations intoned that the degradation of the minidollar would stop via tax cutting, supply-side economics, and a balanced budget — a document that does not include all of the trillions in off-budget debt.

The sorcerers in office today are just as adept as the old ones in spinning the fairy tales of the so-called ''economic growth,'' reduced unemployment, and declining price levels.

Yet, the plain fact is that the current ruler of the United States has inherited what is probably one of the shabbiest currencies in the history of money. The once omnipotent dollar — that was the glory of the globe at the end of World War II — had, by the beginning of 1985, lost 86% of its official pre-war purchasing power and 98% of its unofficial buying power.

17. The gross national product has not risen in 45 years

The public relations people for the US government boast of a 3,711% gain in the gross national product since 1940. Bah. That is pure deception.

In 1940, the gross national product was $99.7 billion. The population was 132 million. So on a per capita basis, the gross national product was $755.

In 1984, the gross national product was 3.8 trillion. The population was 260 million. So, again, on a per capita basis, the gross national product was $14,615. ''But wait!'' shout the public relations experts. ''That's a gain of 1,836%,'' they say. They do not know that they speak in jest.

Using official 1940 dollars, the gross national product is only $2,046 per capita. That is only a 171% increase since 1940. Even that is not a true picture. By my own unofficial statistics, the per capita gross national product is only $438 in 1940 dollars. That is a <u>decrease</u> in the per capita gross national product of 42%.

There has been no progress at all in the United States over these past 45 years.

These facts have never been publicized by the government or the orchestrated press, nor have they been discussed on television. The banking system and the stock exchange community, including its assorted indices, has continued to ignore such details.

As for the generals in the Pentagon, they have been completely indifferent to the deterioration of the American monetary unit which, unfortunately, has destroyed much of what was once Washington's political superiority.

These developments are easily readable from the chronology of budgetary deficits and balance of payments shortfalls in the post-World War II era. They remind me of the dilettantish mismanagement of the pound sterling in the post-war years of 1919-1931, which eventually forced the liquidation of the British Empire.

18. The Russians have more gold than we do

The Russians have more gold than we do. The US owns about 8,600 tons of gold. But the Soviets have more than 9,000 tons.

This information was confirmed to me by a central banker who had a nightly meeting with Soviet Premier Aleksi Kosygin without an interpreter, and they quoted <u>Pick's Currency Yearbook.</u> I don't know if US generals have any monetary knowledge or not, but gold is the most important ammunition in war.

19. The dollar was wiped out once before

America's first currency exchange concerned the old Continental dollar. The Continental was first issued in 1775 with the promise of redemption in hard money.

The promised redemption never took place, and within 6 years, the quotation was 500 or 1,000 paper Continentals for every Spanish-milled silver dollar.

Finally, in 1789, Alexander Hamilton redeemed 100 Continental dollars for one ''new'' dollar. The original issue was over $200 million; less than $2 million was paid out in silver redemption. The US government was the profiteer. This is what is happening all over again today.

20. The US government will crawl into every safe-deposit box to get at your gold

You must protect your future. You must buy gold. Gold prices are low now. It is a good time.

I tell you again, gold is always the single best holding for preserving your assets. Currencies come and go; gold endures. The yellow metal always outlasts government efforts to suppress it. Governments fall. Gold remains — solid and indestructible.

But you must be private about your gold. The governments hate gold. It is too honest. They cannot corrupt it. Something is either gold, or it is not.

Experience has taught me that bank vaults are not accessible to owners of safe-deposit boxes from 3:00 pm Friday to 9:00 am Monday. Devaluations generally take place during this period. Do I need to say more?

Do not store your silver or gold in a safe-deposit box, because there's a chance that one historic Monday morning, the box will have a red seal on it and may be opened only in the presence of government agents. The risk exists.

People forget what happened in 1933, when President Roosevelt nationalized and confiscated gold holdings of Americans. Always remember: the government doesn't like it if you own silver or gold.

The US government will crawl into every safe-deposit box in the country to get at your gold. So keep your silver and gold at home, under the earth. Keep enough above ground so when they come they will have something to confiscate, and they will be happy.

You have had it too good here in America. You cannot believe this will happen here. But it will. No one will prosper. It will be too grim. But those who are prepared will at least survive.

21. You will be lucky if you can take $500 out of the country

The dollar will crash. It will be wiped out. It's coming. It's going to happen here. Maybe even when I'm still alive. At least 50, maybe even 100 or 1,000 dollars will become one new dollar. 1,000 to 1 was the exchange ratio in Brazil.

But first there will be full exchange controls. You'll be lucky if you can get $500 out of the country. Legally, that is. The final step will be the confiscation of gold.

If you want to see what to expect here in America in the future, look at France today. France is the only country in Europe where the krugerrand is allowed.

If you want to travel out of France, you have to have a special currency passport — the same as if you're from behind the Iron Curtain. You can take out 2,000 francs, about $240. That's it.

22. Government loses battle in anti-gold war

During 1980, the seventh year of America's anti-gold war, the United States abandoned its not-exactly-intelligent gold sales. These sales auctioned off gold at bargain prices that have since risen to far higher levels.

For all practical purposes, the world has long since repudiated the so-called ''official'' gold price of $42.22 minidollars per ounce. Washington has not been able to stop the inflationary destruction of the minidollar.

The greenback is incurably ill. And along with causing various political problems during the past decade, this malaise has influenced owners of paper currencies all over the globe to acquire gold.

23. The $12 trillion total indebtedness in the US will never be repaid

I have lived through hyperinflations and currency exchanges in many places. In each case, it was precipitated by a mountain of debt — just like we have here today.

Total indebtedness in the US comes to $12 trillion. It's almost impossible to understand what that means.

There are 260 million people in the country today. Divide your $12 trillion by 260 million. That's $46,154 of debt for each man, woman, and child in the US.

If you want to look just at government debt, the federal government owes $8 trillion. State and local governments owe another half trillion. Total: $8.5 trillion. Government debt.

Does that make you feel better? Divide that by the 260 million people, and it comes to $32,692 of debt for every man, woman, and child in the country. Can you see every person in the country sweating away long extra hours — for years on end — just to pay back this debt? So the US Treasury can continue its crimes against the dollar?

No. There is no chance this debt will ever be repaid.

24. The dollar is doomed

I was in Brazil in 1967 when the cruzeiro was changed over. It was 1,000 old cruzeiros for 1 new one. In Germany, in 1924, 1 trillion paper marks were exchanged for 1 reichsmark. In Hungary, it was 400 quintillion paper pengoe for one new forint, in August 1946.

Nothing has changed today. Argentina already exchanged 100 to 1 in January 1970; then 10,000 to 1 in June 1983; in mid-1985, another 1,000 to 1 was exchanged; and the name of the currency was changed to the austral. Here, in America, I cannot say whether the currency will be exchanged at 100 to 1, or 1,000 to 1. It will

This is a 20 million Bavarian mark note issued in 1923. Ten years earlier, you could have purchased half of Berlin for 20 million marks. Ten years later, it couldn't get you on the bus.

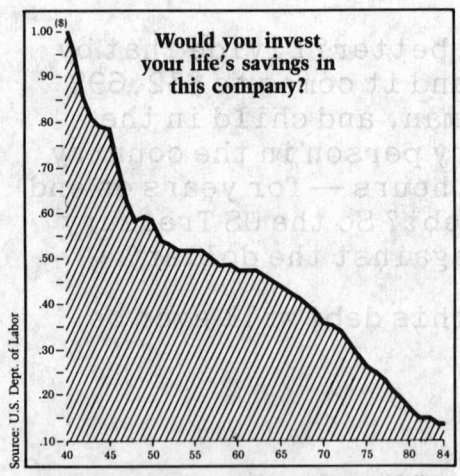

"Of course not! So why are your savings in dollars? This chart is the dollar. It shows—according to actual government statistics—the brutal effect of 45 years of systematic debasement of the dollar."

depend on under-the-table political deals, which I cannot predict.

But this much is certain: The dollar is doomed. As I have said, the dollar is now a pitiful 3 pennies of its 1940 value. One day it will have to crash, as will every unbacked currency which puts its trust solely in God.

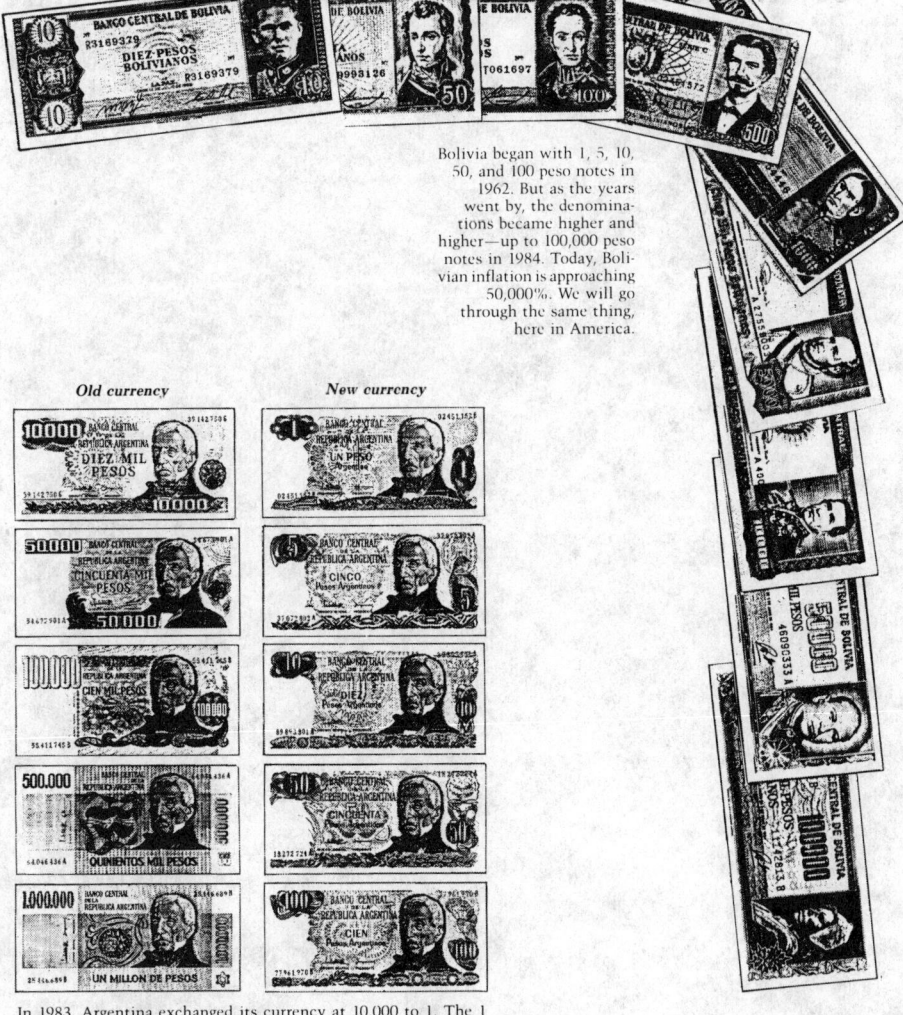

Bolivia began with 1, 5, 10, 50, and 100 peso notes in 1962. But as the years went by, the denominations became higher and higher—up to 100,000 peso notes in 1984. Today, Bolivian inflation is approaching 50,000%. We will go through the same thing, here in America.

In 1983, Argentina exchanged its currency at 10,000 to 1. The 1 million peso, which was replaced by the 100 peso argentino, is worth about 10¢ today.

25. Stamps are not a hedge against inflation

My stamp collection is worth about 250,000 minidollars. But the dealers will only give me half because they want to make a profit on it.

So, there is no hedge in stamps.

26. Inflation has eaten away the golden marrow of the dollar

About 98% of the once mighty United States dollar has passed away since 1940. The dollar suffers from the cancer of inflation.

The dollar was once the world's leading money. It was the main component of the monetary reserves. It was the most respected and sought-after currency on the globe. But a government policy of full employment, backed by high-speed printing presses, ate away the golden marrow of the dollar.

We will create the new hard dollar, and there will be much fanfare and bravado designed to re-establish confidence in the currency. Whether or not that currency already exists, or whether or not the color is red, is of no consequence anymore than its size, denominations, or whose picture it carries.

What is of consequence is that some way will be found — no matter what anti-inflation safeguards are included in the monetary reform package — to get around the controls, and a new cycle of inflation will begin debasing the hard dollar.

27. Real estate is a disaster area

Real estate is a disaster area. In Zurich, I have seen 32,000 cooperative apartments without buyers for years now. May I take bets on Miami and Ft. Lauderdale? You will see 200,000 condominiums without tenants, and this can continue. Overbuilt.

One of my corporate clients told me the following story. The corporation bought a cooperative apartment in one of New York's most stylish hotels for European executives to use when they come to New York to do business with the company. The maintenance fee was over $1,000 a month. The company offered to give the apartment back to the hotel as a present. The hotel refused the offer.

There is a building on Third Avenue here in New York that was built about 20 years ago for $90 million. It was taken over by a trust company in the early '70s for $50 million, and you could have it today for $15 million if you want it. It's the same all over the country.

The only real estate I recommend is farmland and unimproved acreage. The purchase of this property through the short sale of dollars — called long-term credits — has increased in California and the Midwest. It is attracting millions of dollars worth of West German flight capital. Viewed from this perspective, farms worked by their owners, despite huge mortages, will remain a good hedge against all dollar risks.

28. A trick for preserving your estate intact for your heirs

I would like to tell you about a little trick I have picked up.

Every American having assets in Switzerland should have a will with a Swiss lawyer. In his domestic will, the investor should leave a little gift to the Swiss lawyer. This will serve as official notification of death, and the lawyer can then protect the European holdings from American administrators.

This little trick is generally unheard of by even the smartest people.

29. Gold in Switzerland is safe

I recommend you keep a portion of your assets in Switzerland. Switzerland has 100% banking privacy, and the Swiss have <u>never</u> confiscated banking assets. Just prior to the last two dollar devaluations, in 1971 and 1973, you would have gotten 18% more dollars. Furthermore, if an American has money in Switzerland, he will escape the inheritance tax.

In Switzerland, people are completely free to own gold. Indeed, the Swiss franc is backed by gold. That sets a limit on government monetary crimes and makes runaway inflation very unlikely in Switzerland.

Gold in Switzerland is safe. But if you keep some of your gold abroad, don't count on being able to bring it back to the US. As I've warned many times, you can expect to see increasingly tight foreign exchange controls.

Some advisors think that the Soviet Union will invade Switzerland and take it over. They recommend moving your assets out.

Bah. Russia will not take Switzerland. If the Russians attack, the Swiss will fight back. Hard. The whole world would be very angry, and maybe at war.

30. The effect of a currency exchange on dollar-denominated assets

Savings and checking accounts could be blocked under Emergency Banking Regulation No. 1, but probably won't be. However, certificates of deposit might face temporary delays of redemption before new legislation copes with the problem.

Mortgage payments would probably not be subject to any moratorium, in order to maintain the existence of thrift institutions, which would fold if their mortgage investments became valueless.

Bonds and preferred stocks would certainly stop paying interest to their owners, which would not only be individuals, but all kinds of institutions, hospitals, and universities. Annuities, already ridiculous items, may suspend payments or disappear altogether. Life insurance premiums would have to be paid; death benefits would not be affected.

A currency reform can complete the expropriation of all kinds of savings of the population. It can be performed with swiftness and brutality; it can wipe out all public and private bonds, most pensions, all annuities, and all endowments.

31. The most practical gold coins for the coming currency crisis

I recommend the little krugerrands and maple leafs — the one-tenth ounce size. They will be the most practical coins for the crisis you are going to live through.

The one-ounce gold coins may go up to $5,000. That will be difficult to place on the black market. I do not think you would want to get change for that.

32. Governments and corporations prefer to hide the truth

The debasement of statistics is as much a part of the whole scheme as the debasement of the currency. How many people know that the Consumer Price Index has been periodically and systematically falsified since 1947?

For one thing, they fudge the numbers. They are not accurate. For another, they keep changing the starting date of the index. They used to change this starting point every ten years. Now they use 1967 as a base date. That is because it does not look so bad to show that the cost of living has tripled since 1967 as it does to show that the cost of living has multiplied nearly 10 times since 1940.

But nobody wants to know this. The general public, along with the generally unintelligent investor, must be kept ignorant of their willingness to lose most of their assets. The press, radio, and television constantly babble about inflation, yet they do not analyze the problem in terms of constant dollars.

It is nothing but mass self-deception.

33. How to buy gold in London

Investors who have substantial sums should buy the 400-ounce bars. Buy them in London, if possible. If they do not leave the warehouse, they will not have to be assayed. Then, these bricks of solid gold, fully paid, can be sold within an hour, and give more protection than any kind of bond or stock.

The smaller bars have a drawback. A one kilogram bar, which is some 32 troy ounces, cannot be sold until it has been assayed. And all the little assay shops here are crowded with orders. Sometimes it takes three months before you can get the piece assayed.

By the way, when World War II broke out, I had 2,600 sovereigns in a London bank. After the war, they were still there. Gold owned by non-residents has never been confiscated in England.

If you are interested in little recipes for dealing in the gold market, may I suggest that you buy a 400-ounce gold bar in the manner I described and then ask the bank to use that bar as collateral for the purchase of 3 additional bars in the forward (futures) market. This way you have gold outright and a leveraged gold position, as well.

34. There will be two currency exchanges: the first will be benign, the second malevolent

There will be two currency exchanges. The first, which will be benign, has already been announced by the US Treasury.

They say the new currency is to prevent counterfeiting. They claim that in a very short time, the technology will exist to copy the current notes, and it will exist on a wide basis.

This may even be true. Not every word out of their mouths is a lie.

I realize that many people are upset and worried about this new currency design. But I'm very old, and I have perspective that many young people cannot share. New redesigned currencies are issued all the time, for reasons that are harmless. They circulate, side by side, with the old notes for a period of time specified by the government. Gradually, the old notes go out of circulation, but still remain legal tender before eventually losing this status.

I don't think honest individuals who have stored cash will have much to fear from this currency redesign. Most people in this category have amounts of a few thousand dollars, and that will be no problem.

If you live in an area with many banks, it's a safe bet that you'll be able to visit many of them and exchange, say, $1,000 to $2,000 at a time without arousing suspicion.

If you change $2,000 per visit and visit three banks a day for a week, you can change $30,000. Most of you probably don't have much more than that. I'd suggest you go to banks where you aren't known so you will not be conspicuous.

This currency redesign is NOT the currency exchange that I have been predicting will come to the US. That will come later. You should worry about this currency exchange; it will be malevolent. I believe 100 old dollars will be exchanged for 1 new one. But it could be even 1,000 or 10,000 to one. It will be a time of great suffering.

35. American currency policy since 1940 has been nothing but an unbroken series of flops

The present administration is helpless. They don't know what to do and they take any stupid advice they can get from whatever overambitious people happen to be in favor at the moment.

No, the US government is not adroit enough to master the currency problem. There is not an institution of higher learning in this country teaching a full-time course in currency theory. So how can the civil servants who run the show for the nation learn anything? Will Mr. Volcker succeed with these measures? I say no. He is intelligent, but he is against gold, and he cannot cure a currency which has lost, unofficially, 98% of its purchasing power since 1940. He cannot bring it back to life.

I am sorry to say the situation is hopeless, no matter what Mr. Volcker does. The inflationary process has its own dynamics and the pattern is always the same. During the destruction of the Austrian korona, the Czech koruna, the German mark, the French franc, and the British pound, the process was the same as in the United States today, with intermittent deflationary trends along the way. The French franc went through numerous devaluations — maybe that is in store for us, too.

American currency policy since 1940 has been nothing but an unbroken series of flops. By the end of October 1979, the unofficial purchasing power of the greenback had shrunk from 100 cents to only seven. Today, it has shrunk to about 3¢. This destruction and irreversible rise in the cost of living has been hidden by a deliberate lie called the Consumer Price Index. Most of those who own imaginary paper assets, including cabinet members, presidents of blue-chip corporations, and pension fund managers do not want to know about the decomposition of what they believe they own.

So you see, I have little hope that Mr. Volcker will be able to single-handedly correct such a bad situation.

36. The only gold holding that can be kept in your safe-deposit box

I am afraid that one day the government will indeed call gold in. Gold bullion will be subject to confiscation.

This is one big advantage to numismatic gold, such as the double eagles. It is an idiosyncrasy of governments that although they may prohibit ownership of gold in any form, they are reluctant to touch collections of numismatic gold coins.

Today, there are some 49 countries which forbid ownership of gold by their citizens, but do allow holding gold coins for numismatic purposes. Even the Soviet Union and Eastern European countries legally tolerate the acquisition of numismatic gold coins. So these are the only gold holdings that could be kept in your safe-deposit box without any fear of confiscation.

37. Our assets are melting away before our very eyes

The budget deficit is currently running at about $200 billion per year. This is financed through monetization of the debt. The Treasury will issue more bonds and the innocents, especially the banks and the Federal Reserve System, will gobble them up.

But this process increases the money supply, thereby eroding the purchasing power of all existing minidollars, and the result is higher prices. As we inflate at higher and higher rates, what happens? The dollar melts in your hands.

One of the most shocking accusations I have for you is that, while we may be the richest nation in the world, our assets are melting away before our very eyes, and virtually no one recognizes this fact. In truth, the great industrial wealth of the United States is a sham and a delusion.

38. Few people understand the concept of currency debasement

This process of debasing the currency to pay for government deficit spending has been going on for centuries. The Egyptians did it, the Greeks and Romans did it. Countless other nations have done it. Now it's going on all over the world. The process of monetary inflation — and its result, soaring prices — is a simple concept. Adam Smith understood it, as did John Stuart Mill, David Ricardo, and other classical economists.

But, alas, today few people understand the concept. Instead, thanks in large part to the writings of John Maynard Keynes, higher prices are laid at the feet of excessive labor wage demands, greedy corporations, Arab oil sheiks, and the disappearance of anchovies off the coast of Peru. *Mon Dieu!* The media — woefully ignorant of currency theory — propagandize these stupid explanations, and the public is left totally in the dark as to the real cause.

39. International transfers of currency will require a special license

Unstoppable inflation, balance of payments and budget deficits, rising silver and gold prices — these signs tell us that the currency reform process is already underway. After seeing the futility of IMF and US gold sales, the administration is considering controls for the economy in the faint hope that this will stop the decline of the greenback.

These plans center around wage and price controls, and lower taxes. The creation of such controls will immediately lead to ''unsupervised'' wages and prices, and the controls will be a flop — even though the public today favors such measures.

As the dollar continues to drop, the amateurs in various government departments will demand stronger controls with heavier penalties for disobedience — Iron Curtain style — and then, finally, they will propose foreign exchange controls. International transfers of capital would then be allowed only under special license, foreign currencies would be surrendered, gold would be nationalized, and the futures markets in precious metals would be closed. This would only transfer the trading activity to Canada, England, Switzerland, Singapore, and Hong Kong.

One or more foreign countries would stop accepting US dollars. There would then be a statement declaring the US dollar in better shape than ever, and Washington would freeze foreign assets. But all rescue schemes by Treasury officials or central bank executives will not be able to save the dollar, except by an official state bankruptcy, for which tranquilizing words would have to be found by the public relations artists.

We would then see a ''temporary'' moratorium on all debts, as well as international payments, the creation of a new currency unit with some link to gold, and the issuance of new bank notes against the old ones. Keep in mind that although these conditions do not seem to be in the near future, they could be triggered by the refusal of any country to accept dollars in exchange for its currency or products.

40. Our "conservative" president will not be able to stop inflation

During the current president's reign, there have been the usual press releases about alleged financial and economic improvements, with their usual repercussions on the futures markets.

But what no one will dare admit in public or private is the unassailable fact that the greenback has to be debased into nothingness — thus reducing the value of all debts to zero.

For it is only by continued monetary and price inflation that the American economy can function.

As a very astute former governor of the Banque de France observed, an annual price inflation of at least 10% — as reported in the official statistics — is needed in order to keep a financial and economic system liquid enough to work.

These Himalayas of massive debt and bank credit can only be climbed with ever-higher piles of paper — paper that has nominal value and rotting purchasing power.

And in the philosophy of inflation, more can only mean less! The $185.3 billion US budget deficit for fiscal 1984 was the second highest ever. But when it's adjusted for inflation, in real terms it's a large drop from the record $207 billion shortfall in fiscal 1983.

America will continue to live under the pressures of inflation in order to keep the economy, as well as the financial markets, functioning. Destruction of the country's purchasing power continues. And prices will keep on rising in the supermarkets, in the hospitals, and around the rings where silver and gold are traded.

41. Bah. We will not have deflation in the United States

A growing number of people seem to believe the US economy is heading for a deflationary crash. Bah. We will not have deflation in the United States.

As you can observe by looking at the official US cost of living statistics, the dollar has been debased to less than one-sixth of its 1940 value. My own unofficial figures put the value of the minidollar at around 3% of its pre-war value. Monetary history shows that any unit that loses more than 25% of its purchasing power is ripe for devaluation. The minidollar is overripe. But there is little fat left to be melted down.

When the purchasing power of the 1940 dollar has been reduced to about a penny, it will mean that just about all the capital has been used up, that government debt has been wiped out, that all paper assets have been destroyed, and the game will be up for this inflationary cycle. As you can see from the statistics, that point is not far away.

We will then be faced with 2 choices. One is deflation and depression much like the 1929 to 1940 era in which the slow process of savings and capital rebuilding takes place. The other alternative is a currency reform. I do not believe we will have deflation. It need not happen. We need a new currency, and I believe the old dollar will be replaced by a new ''hard'' dollar. We are going to change 100 old dollars to 1 new hard dollar. I believe it will take at least another 12 to 24 months before all the assets have been debased, and there is nothing left to devalue.

Deflations do not win elections. To liquidate America's debt structure by deflation would be political suicide for the party in power. The only other alternative is to have the debt repudiated by further debasement of the currency, through another massive creation of debt of colossal proportions.

With at least $28 of credit required to create one dollar of the future _real_ gross national product, titanic infusions of money and credit will be needed to reactivate economic activity. Although the central bank will begin the process slowly, it will find that more and more will not be enough.

What will be needed is the inclination of entrepreneurs and the public to take on more debt — an activity they will shun for quite awhile. The reason: They are at present too worried as to how they will meet their current financial obligations, let alone future ones.

New debt creation will not begin until — and only until — corporations and households in the United States are subconsciously convinced that such new debt will be repaid with further debased minidollars. Or better yet, microdollars.

Once again, currency reforms are not new. Austria, Hungary, Germany, Poland, and Greece were some of the countries that had currency reforms after World War I. In recent years, several nations — all faced with economic disaster because of currency debasements —

adopted successful currency reforms and stabilization policies, and enjoyed economic recoveries. West Germany in 1948 exchanged 10 old marks for 1 new mark. In France it was 100 to 1 in 1960. I was in Brazil in 1967 when they exchanged 1,000 old cruzeiros for 1 new one. Argentina traded 10 for 1 in 1970, and was successful for awhile. The currency reform under DeGaulle and the exchange of 100 old francs for 1 new franc saved France.

42. Sooner or later all monies die

Let me tell you a story. On March 29, 1984, the British government ceased minting halfpennies.

Originally, they made pennies out of silver. They had a Christian cross on the front. The penny was worth so much, that they made the line of the cross into a groove. When you bought something, they could break the penny in half, if need be, and give you change. That was 700 years ago. I am not that old, but I studied monetary history when I was a young man.

When they started minting actual halfpennies, in the 13th century, they were also made out of silver.

Through the years, always, they made the halfpenny a little smaller. A little less size; a little less purity. Then finally, copper squeezed out silver. That was in the 19th century. Today, the halfpenny has disappeared entirely. The fate of the halfpenny may seem insignificant, but that is the fate of all currencies.

In Spain, for example, there was a long-lived currency, the escudo. At the start, it had almost exactly one-tenth of a troy ounce of gold. By the middle of the last century the escudo had been so debased, it was replaced by a new currency, the Spanish peseta.

The peseta, too, started out as a gold coin. That was in 1870. By today, it has lost 99.8% of its value. The peseta — which, again, started out as a gold coin — is so debased, you need 8 to make up one nickel.

In Spain, they do not even recall the escudo, except as the currency of their neighbors in Portugal. Monetary amnesia. It is unfortunate, but they will have to go through the wringer again.

How do currencies die? Some of them fade away rapidly. Some fade away slowly. Some are wiped out following financial calamities. Sometimes the nation collapses and the currency disappears, as here in America in the southern Confederation. Other times, the currency disappears and that destroys the nation. The fate of the nation and the fate of the currency are one and the same.

Sooner or later, all monies die of wear and tear as a result of clipping or reduction in weight, or of debasement, depreciation, or devaluation.

43. The danger in opening a safe-deposit box under an alias

In Canada, they will not confiscate gold because they have a gold mining industry.

However, some years ago, I had a bad experience with gold in Canada that may prove instructive for you. A few of my clients died, and the children came to me with the keys of safe-deposit boxes located in Canada. They were taken out in false names.

Let us suppose the person who died was Mr. Jones. He had rented the safe-deposit box as Mr. Green. So the heirs couldn't get in. The gold was lost. It may still be in the same small box.

44. Since 1973, the value of worldwide gold hoards has risen 301%

The worldwide total of privately-owned gold is over 26,000 tons. This exceeds the combined official gold reserves of Austria, Belgium, Canada, France, West Germany, Italy, Japan, the Netherlands, Portugal, South Africa, Switzerland, the United Kingdom, and the International Monetary Fund.

To produce this quantity of bullion, at the present annual rate of free world production, would require 27 years.

Since 1973, worldwide gold hoards have risen in tonnage by about 8%. At the same time, their value in official constant 1940 US dollars advanced 75%. In terms of minidollars, the gold's value climbed 301%.

This performance in the preservation of purchasing power cannot be equalled by any paper money, or by any government or corporate bond.

45. Fully paid holdings of gold bars and coins should not be sold

The outlook for gold cannot be regarded as a dim one. The various pressures, mostly psychological, which are working against the yellow metal will eventually subside — but not the dynamics of inflation.

It is the greenback which is unstable, and not bullion. Double-digit interest rates are fleeing from American money markets. And since all economic activity in America is geared to the debasement of currency, fully paid holdings of bars and coins should not be sold.

The present political and monetary decay is something that cannot be undone soon, if at all, inasmuch as the gigantic graft and corruption in America wishes it to go on as before for bigger and better tax-free profits — and will continually frustrate all attempts to control the money supply.

The omens are not good. To master inflation seems beyond the reaches of the men in power, just as the mastery of gold has eluded all their predecessors since 1940.

46. The stock market is a graveyard — but they refuse to bury the bodies

No one recognizes what is happening because we continue to use and work with numbers that are absolutely phoney and useless. Almost none of the figures we use are adjusted to reflect the impact that inflation has produced over the past 45 years.

The stock market, for example, is a graveyard, but they refuse to bury the bodies. The constant dollar value of these stocks is nowhere near what their apparent or nominal dollar value appears to be. The Dow Jones Industrial Average is NOT around 1300. It is around 180 if your deflator is the official government cost of living index, basis 1940.

My own unofficial calculations place the Dow at about 40. But stock market analysts, who have been wrong for so many years I do not wish to even count them, continue to look at the apparent market value of stocks through their rose-colored glasses and make their asinine predictions that are dutifully spread throughout the land by the press.

Should we become nasty and disrespectful, we would have to dig a bit deeper in our analysis of Washington's and Wall Street's pseudo-truth and discover that the IBM share is worth not more than $3 in unofficial constant dollars. Of course, this figure would not be believed by brokerage houses here or abroad. To print them in the prospectus of a new stock issue or in a promotional folder for some investment trust would not meet with the approval of the Wall

Street moguls, and I do not think they would appear in the announcements of a new government bond issue.

If you want the truth, you must adjust the market for inflation. I did this some years ago, using only <u>official</u> US government Department of Labor cost-of-living figures, and the results are plain for anyone who cares to know. When I first pointed this out to French president Valery Giscard d'Estaing 12 years ago, he said, ''But my dear friend, since 1965 there has been no progress whatsoever in the United States.''

You will note that the trend continues.

47. The history of currencies is not sympathetic to men who tried to stabilize the money with deflation

The economy and currency of the United States are now enslaved to the tyranny of inflation. It is the despotism of debasement which rules the land, and to which the political liberties will have to be sacrificed.

Like some refined torture of the Renaissance, the fortunes of America have been strapped to the powerful stallions of inflation and its twin — deflation — which will soon be dismembering the social order.

At present, the ruler in Washington clings doggedly to his scenario of economic and monetary ideology, which is to reform the currency morals and habits of the country. His perceptions of disciplining a generation accustomed to the outpourings of inflationary money could easily bring him a very unhappy fate.

The history of currencies is not sympathetic to men who tried to stabilize the money with deflation. During the French revolution, enemies of the Assignats either fled the country or were put to the guillotine.

In more recent times, such opponents of the inflationary process have met with death or political exile. In 1919 in Czechoslovakia, the Minister of Finance, Dr. Alois Rasin, tried to stabilize the country's disrupted currency by means of deflationary policies. For his courage and integrity, he was assassinated by a civil servant who had been discharged as a consequence of cutbacks in the payrolls of the Czech bureaucracy.

During 1931–'32, the German Chancellor, Heinrich Brüning, and the President of the Reichsbank, Dr. Luther, attempted deflation to restore the solvency of the government. Both were shortly replaced by Adolf Hitler.

Any attempts to try and stabilize the US currency with deflation, today, will meet with the same failure.

48. I recommend krugerrands to protect you from my other clients

I recommend the krugerrand over other coins to protect you from some of my other clients. One of them has a private mint in Bombay where he makes sovereigns. His sovereigns look good, but they have less gold than the real ones. There is another mint in North Africa making sovereigns.

And there are people in Damascus and Singapore making double eagles. The Mexican peso is also being counterfeited.

The only safe coin that you can buy is the krugerrand. Although the technical perfection exists to counterfeit it, it is not being done. Why? Because the premium is too low to make it worthwhile.

Canadian maple leafs can also be considered safe from counterfeiting. Their 24 karat purity makes it difficult, and here again, the low premium makes it unprofitable.

49. If all the silver in Washington's stockpile is sold, the profits will not wash away a speck of red ink from the budget

As a small footnote to history, Washington has carried its silver stockpile on its books at a value of just 90¢ per ounce.

Nevertheless, even if all the silver is sold, the profits generated at the higher market price would not wash away one speck of red ink from the budget.

50. How a paper money panic affects investment demand for gold

In 1974, panic, sparked by a flood of currency devaluations, brought on a new gold rush. Besides the downward floating French franc, the 11 countries of the Franc Zone saw their monetary unit, the CFA franc, drop.

Spain set her peseta free to depreciate, as did Greece with her drachma. The gold content of the shillings of Kenya, Tanzania, and Uganda was cut, while the kwacha, the money of Malawi, was turned into a floating currency.

Japan *de facto* devalued the once-mighty yen. Australia's dollar, after having been upvalued several times, was being groomed for devaluation. The Hong Kong dollar was also being talked about as a devaluation candidate. As for the British pound sterling, now a shabby micro-unit, it plunged to US $2.16 — an historic low.

Worldwide, the public rushed to buy gold. Americans, in flagrant violation of Washington's comic book prohibitions of gold bullion ownership, lined up at Canadian banks to buy gold.

Gold smuggling by the famous passeurs from Switzerland into France, which forbids the export and import of the yellow metal, rose substantially. Singapore fed the gold-buying hunger in Japan. Gold fever swept through black markets in Iron Curtain countries.

Hong Kong reported a mad rush to buy freshly imported South African krugerrands. Dealers were cleaned out, with sales in the thousands.

51. The United States came close to a total financial collapse in early November 1978

During the four years that the Carter Administration was in office, its shabby abuse of power in an attempt to eliminate the yellow metal completely from the minds of man ended in miserable failure.

During the 48 months of horrible mismanagement of the currency, the regime could only reinstate sales of gold bullion as well as talk down the minidollar in foreign exchange markets.

The public has yet to realize how close this country came to a total financial collapse in early November 1978. The relatively hard currencies of the world, as well as silver and gold, were then soaring to previously unheard-of heights against the minidollar.

Anguished governors of foreign central banks tried to reach someone — anyone — in Washington with some responsibility to avert the impending disaster. Only a last-minute rescue operation saved the minidollar from plunging into premature oblivion.

Washington was finally forced — by Switzerland, West Germany, Japan, and other foreign governments — to support the dollar.

These nations had to aid the ailing currency in order to prop up the value of many millions of minidollars held as reserves in their own central banks.

Since Washington did not see fit to stop the dollar decay at that time, the advance to double-digit inflation generated hefty new petroleum price increases by the OPEC powers.

And because the increased gold sales by the Treasury could not turn back the rise in price of the yellow metal, the gold auctions have been suspended indefinitely — hopefully forever, if the current president wants to regain any respect for the greenback, which was once the world's most treasured monetary unit.

52. The new paper currency will be tied to gold

Silver and gold will play an important role when the currency is exchanged. My guess is they will try to rebuild the past and go back to gold at $35 an ounce after the state bankruptcy.

But the 35 new dollars will correspond to many, many more minidollars.

We will have silver coinage again. The new paper currency will be tied to gold.

We cherish corruption. Nothing will ever change. Administrations come and go. Governments expropriate the people through the debasement of the currency, but gold always triumphs over the embezzlers.

53. If the true condition of the US corporation were known, no one would buy stocks or bonds

There is no longer any real industrial progress in the United States, you see, because there is no capital. The silent, constant rape of the dollar through the inflationary process hollows out the assets of the corporations the same as a worm hollows out the insides of an apple until only the skin remains. Do not blow too hard upon the balance sheets of the major corporations and banks in the United States. You might make them collapse.

But corporate presidents and their comptrollers don't want to talk about reality in their balance sheets. They prefer the luxury — and illusion — that the use of nominal dollars offers, compared with the realism of constant dollars. Recently, I took 5 of the top companies, including 1 bank, and applied constant 1940 dollar deflators to their balance sheets in order to remove all the illusions of deflation. Here are some of the results of my findings...

The president of one of the largest companies in the world could proudly announce to his shareholders that dividends on common stock, including splits, had increased 149% since 1940. That is the illusion. Now for the reality. In constant dollars during that same time period, the company's dividends <u>decreased</u> in purchasing power by 39%. Worse yet — as if that isn't bad enough — the decline was 76% if you use my own unofficial estimates of the rate of price increase, which I consider to be closer to the truth than the figures issued by the government.

Let us continue our stroll through this cemetery. Here is a major chemical company that thinks its net income has increased 428% since 1940. Nice work, you say? Not so. The increase was only a meager 30% using the official constant dollars, or a decline of 48% if you apply Pick's unofficial deflators.

Here is one of the world's largest steel companies that will tell you its net income increased 302% since 1940. It is a pity they do not realize their net income <u>decreased</u> 1% in 46 years using <u>official</u> government statistics. My unofficial figures indicate their net income decreased by 60%.

Total capital funds of a major New York bank did not increase 288% in 46 years, but fell 5% using official statistics or plunged 62% by applying my own handiwork. Should we say a requiem mass?

Some of these corporate executives are aware that what I've been telling you is the truth. But they wouldn't be caught dead talking about it anywhere in public in the United States. I know this is true, because some of them come to my London seminars and tell me so. I don't blame them. They are but pawns in the government's game of inflation.

But none of this is new. Toward the end of all great inflationary cycles, businessmen begin to discover that all of the sales, profits, capital, and reserves they thought they were enjoying had suddenly vanished.

54. A loaf of bread will cost $100

The growing invisible debt will be the pacesetter for a new inflationary cycle. We can continue this for another two or three years.

My personal indicator is a loaf of Pepperidge Farm bread. When I started watching the price many years ago, I think it was 19¢. Now it is over a dollar.

It is hard for you to visualize a world in which a loaf of bread costs $100. But it is coming. I give you fair warning. One year at Harvard will cost $1 million.

They will stop taking $1 bills — too much bother for something so worthless. They will have to stop dealing in hundreds and thousands and get used to millions and then billions, and then, finally, trillions. When the currency exchange comes, it will be welcome.

I have no timetable. I did not study astrology. But the pace of events is quickening. I recognize the signs. We are getting closer.

55. Since 1933, the US budget has been an instrument of electoral bribery

Politics, in an age of inflation, are made in the streets. It will not be difficult for the politicos, of whatever persuasion, to find the necessary rationale to wipe out more of the greenback's purchasing power.

Politicians are fundamentally overgrown adolescents with too much self-esteem, too little self-respect, and absolutely no shame. Through their artifices of deception and duplicity, they build castles in the air — and then try to live in them! In non-political terms, such personalities are usually found as inmates of mental institutions and are generally known as schizophrenics.

No one in Washington wishes to be reminded of the fact that the budget of the United States has been, since 1933, an instrument of electoral bribery. Its costly social programs and subsidies to business — especially the agricultural and housing industries — have made them the spoiled brats of this country's economy.

56. How to buy gold in Switzerland

You should keep some of your gold in Switzerland.

One way to do this is through the gold warehouse receipts issued by Mocatta Metals Corporation, which represent direct ownership of physical gold — bullion or coins — in specified warehouses situated in tax-free locations in Switzerland (or Delaware, if you wish domestic storage). They sell for a 2% or 3% premium over bullion. Storage cost is about 0.5% per annum.

These warehouse receipts, signed by the bullion dealer, warehouseman, and insurance company, are a very convenient way to hold readily transferable gold. They are fully acceptable for payment of gold obligations to any major recognized bullion dealer.

Mocatta warehouse receipts are available from C. Rhyne & Associates, 110 Cherry Street, Seattle, WA 98104, (800) 426-7835, (206) 623-6900 OR Western Federal Corp., 8630 E. Via De Ventura, Western Federal Bldg., Scottsdale, AZ 85258, (800) 528-3158, (602) 998-1000.

I do NOT recommend the gold certificates or gold passbook accounts issued by some large American banks. The American banking system is a fawning vassal of the US Treasury.

57. Inflation under 10% is merely an illusion

Politicians through the ages have known that the voters do not want realism; they want magic! And it seems as if the present ruler and his technicians fancy themselves as prestidigitators *par excellence*.

Stopping inflation, or even reducing it below 10% a year, is a political illusion. So long as all the so-called ''off-budget'' borrowing by the various virtually autonomous agencies remains practically untouched, then it is an act of hypocrisy to believe that the finances of the government can ever be sanitized.

Nonetheless, optimism of almost messianic proportions greets the speeches from the throne. The ruler told us that taxes would be cut to activate a fresh avalanche of goods into the marketplace — and this will meet the so-called inflationary demand supposedly resulting from too much money chasing too few goods.

But the automobile showrooms remain filled with vehicles that cannot be sold, while residential housing found few buyers.

As for increased productivity reducing the rate of currency debasement, how much more hair can a barber cut, or how many more shoes can a cobbler repair, or how many more pastramis on rye can a delicatessen put together?

58. 40 years of dollar debasement cannot suddenly be reversed

The road to the record decline of the greenback was gilded with all possible public relations techniques explaining to the public at home and to the world that whatever Washington did was good for all the people on this planet.

The object of this propaganda was to tranquilize all the capitalistic and patriotic holders of the $1.7 trillion of government debt certificates issued to pay for Washington's expenditures that could not be covered by receipts.

It goes without saying that all these shortfalls were covered by the printing press in the form of engraved paper. To reduce public fear, abundant monetary sugarcoating was applied to harsh facts with masterful virtuosity.

The official cost-of-living index was changed at least once every ten years to deceive the public and prevent them from having qualms about any kinds of bonds. It was always a policy of sweeping past blunders under the carpet of history. In the meantime, the money supply rose and rose.

The result is that America is wholly dependent on the creation of more debt and more inflation to keep an economic contraction from snowballing into a depression of gigantic proportions.

Nowhere is there mentioned by the regime what is *sine qua non* of any attempt to bring a halt to the debasement of the monetary unit — currency stabilization, in terms of foreign exchange value as well as purchasing power.

As long as this most fundamental thought remains absent from the thinking of the managers of the minidollar, whatever new edicts come from the occupant of the Oval Office will meet with little success.

Debasement of the dollar has become the indispensable means of pushing the economy to greater increases in so-called ''growth.'' The inflationary process, after 40 years, cannot be reversed by suddenly embracing monetary puritanism.

59. Get some of your gold out of the country NOW — before it's too late to do it legally

I am too old. They cannot harm me anymore. But you, my young readers, must protect your future. You must buy gold.

I tell you repeatedly, I have seen this before. Gold is always the single best holding for preserving your assets. The yellow metal always outlasts government efforts to suppress it. Government falls. Gold doesn't. People know that gold will always protect its owners, whatever governments decide.

But gold is not enough. You must be very private about your gold. Keep your silver and gold at home, under the earth. Keep enough above ground so when they come, they are happy. They think they have gotten your gold.

And get some of your gold out of the country NOW — before it's too late to do that legally.

Of course, owning bullion coins after the government tries to confiscate them will be illegal. America's underground economy is going to expand further than you can believe. It will be something like Russia, with the most severe penalties for the so-called economic crimes. These will not be crimes as you and I know them, but the so-called crimes of innocent men and women tryng to protect their financial lives.

You will see street vendors in midtown Manhattan and on Connecticut Avenue in Washington. They will sell you whatever you want and take your gold in exchange. The price they offer you will reflect a discount, but they will not report you to the IRS. No questions asked.

At some point, gold will become legal again. Those who can go to their garden and dig out 10 ounces of gold will be much better off than those who can't.

I know you cannot believe this will happen here, but it will. Those who are prepared will survive. No one will prosper. It will be too grim.

60. Two metals which are underpriced

I advise my clients to buy platinum, because it is underpriced. I believe platinum should be $100 more than gold. When it is less than that, it is underpriced. Copper, also, should be higher than it is today.

Titanium and other strategic metals are beautiful on paper, but you cannot sell them. You will be cheated in the end.

You have to buy something which has daily volume, like copper, gold, silver, and platinum. I advise my clients to buy precious metals and sit on them. The posterior is very important to the preservation of wealth.

61. Silver will not be confiscated when they come for your gold

Silver will do very well in the future. Not nearly as well as gold, but very well. I am convinced that one day it will go back to $40.

However, for the time being I'm nervous about silver. Until recent months, when the price dipped below the $6 level, the short sellers and dealers appeared in the futures markets and drove the price down. Now it is around $6.10. Will the short sellers disappear from the market? Will Kodak invent a film that doesn't use silver? Silver should go at least to $10. But will it? I don't know.

One feature about silver that should interest you, however, is that it will not be confiscated when they come for your gold. That's because it is an industrial and strategic metal, and because we cannot have a silver currency.

62. When the currency exchange comes, the suffering will be terrible

The currency exchange and devaluation officially acknowledges that the public's money has to be expropriated. It also means that the government has cancelled its debts. If you can't repay it, cancel it. People who have debts will win. People who have savings accounts will weep.

The value of government bonds are already illusory. With the dollar really worth only 3¢ today, a $1,000 government bond is really worth only $30. $970 of the value has been expropriated. With the currency exchange, even the illusory value of these debt certificates will be wiped out, and the bondholders will get little compensation for their losses. And at least half of the US population owns bonds, either directly or indirectly, through bank savings accounts, insurance, or mutual funds.

With all the adaptations to the new monetary setup, we'd have a temporary decline of the economy. We'd see sudden decline of purchases by consumers. Corporations — manufacturers and retail stores — whose cash holdings are not large enough to pay wages in the new currency, will go bankrupt. Many people will lose their jobs. The suffering will be terrible.

63. The minidollar is ripe for a ruthless liquidation

As the financial situation of the United States deteriorates, early signs of an involuntary deflation have begun to appear. Inasmuch as the current minidollar has less than 14¢ of buying value left in terms of official constant 1940 US dollars, and is worth 2¢ to 4¢ in terms of unofficial constant 1940 US dollars, the onset of the terrifying force of a full deflationary wave would unleash social as well as financial havoc.

The multi-trillion dollar public and private debt pyramid rests on a minidollar without substance. It has — as the Germans would say — lost almost all of its *Substanzwert*. It is ripe for a ruthless liquidation.

For that is exactly what deflation means — the paying off of debts, mostly via bankruptcy. That is the paramount problem with which the ruler of America should occupy himself, and not whether the budget will be balanced or not, nor what spending cuts should or should not be made.

The hundred billion minidollar deficits that will be generated by the regime in the years to come are puny in comparison to the multi-trillion dollar debts that will have to be repaid in one way or another.

If the public and private debt of America had been held to modest proportions relative to the output of goods and services during the post World War II era, a cyclical downturn in economic activity would only be a temporary interruption in the flow of real — and not paper — prosperity. And a currency of stable purchasing power would be guaranteed. But for the last four decades it has been easier to embrace the inflationary process, so as to buy votes.

64. People believe they can live with inflation forever — but one day the bubble will burst

During the more than 5 decades of the current monetary inflation, America has come to love what is really the pursuit of a pot of gold at the end of the rainbow.

The orgies of paper money issued from Washington since 1940 have triggered numerous bull markets on Wall Street — so the financial community (and investors) adore it. Businessmen have loved it too, as it inflated paper profits.

Homeowners have succumbed to it, finding themselves becoming rich on paper. This became a painless process, as they were paying off mortgages with cheaper greenbacks. Meanwhile, the labor unions have shouted endless ''hosannahs'' as paychecks became fatter and fatter.

Like Peter Pan — the boy who never wanted to grow up — the American public has come to believe that everything has to keep going up. Except, of course, the cost of living. A rising Consumer Price Index is not part of the happy scenario.

This is why most people don't give a damn about inflation or what happens to the dollar.

In Israel, the increase in the cost of living is over 350% per year, and people live with it. In Bolivia, inflation fluctuates between 10,000% and 50,000% per year, and people still live with it.

Here, it is not yet so bad. But our time is coming. The storm clouds around the dollar are darkening. People believe they can live with inflation forever. But one day the bubble will burst. The dollar will be wiped out. Kaput.

The contracting economic cycle has already begun and will perhaps end in a full-scale depression. This will, of course, increase the present unemployment figures and also lead to more inflationary measures by governments. It will force not only the United States but also most other industrial nations to proceed with devaluations and drastic currency reforms in order to put people back to work. I now believe it may be several more years before Washington is forced to create the new ''hard'' dollar. You see, a currency reform is nothing but a fancy name for state bankruptcy.

When that happens, the minidollar will be absolutely and totally wiped out; it will be finally buried and will take into its paper tomb all dollar debts. Such a worldwide repudiation of debts will lead inevitably, despite the opposition, to a currency reform, since all assets and liabilities must be defined in the new currency. The debts of the corporations will be adjusted to the factual level of the debased minidollar. Most likely, they will simply be written off according to the amount of their outstanding bonds.

Top officials all over the world seem to agree that the dollar will continue declining until it becomes valueless, to be replaced by a newly created bank note. The exchange of bank notes into one, new ''hard'' dollar will wipe out all debts, official and private.

65. Inflation has made "investment" a pornographic word

Today, there is no such thing as investment. The word means to place money for a return. But there is no longer any real return — only the illusion of return that is created by inflation. Inflation has made ''investment'' a pornographic word. Today there is only gambling.

We have walked together through the graveyard of corporate balance sheets and the stock market. Where is the return? We have looked at government and corporation bonds. The yield is what...6%, 8%, 10%, whatever? Assemble the inflation rate of, say, 10%, mix in the taxes due on the euphemistically-designated ''unearned income,'' and what do you have? You have a negative yield.

66. There are only two sensible uses for diamonds

Diamonds are a dealer's best friend, not a girl's. They are a good business for De Beers, but not for you and me.

D-flawless diamonds have come down from $100,000 to $30,000 a carat. If you buy a diamond today, you will have to sit on it for at least five years to get your money back because the mark-up is so high and the buy-back so low.

Diamonds are basically to be worn and to give pleasure to women when they reach a certain stage in life. They may like them, but they are not the greatest inflation hedge.

67. I will never go short gold, nor will I ever be out of the gold market

Gold has outperformed the stock market and the bond market. If we have a new currency, we will again begin inflation, and gold will be needed. I will never go short gold, nor will I ever be out of the gold market. I do not see any opportunity of ever selling one ounce of gold.

Let me give you one humble bit of advice on how to make money in the gold market. Buy gold and sit on it. That is the key to success. Against my advice, a client of mine fooled around in the gold futures markets and lost $350,000 — of his mother's money. Sit on it.

Sometimes I get discouraged when I think that only about 2 million people of 260 million Americans have the vaguest idea of the concepts we have discussed.

Yet I am encouraged now and then that my educational effort is progressing. May I tell you a personal story to illustrate? Some time ago my Ilse announced that her Oldsmobile was aging, and she had to have a new one. It's a legitimate problem in every household. So I bought her a kilo bar of gold that at the time was worth $3,000. I told her that with her Oldsmobile and the kilo bar of gold, she could buy a fine new automobile. I am pleased to report that she still has the Oldsmobile — and the kilo bar of gold.

68. The "Great American Inflation" is the result of willful abuse of the creation of money and credit

The ''Great American Inflation'' is not the result of too much demand and too little supply. It is the result of the willful and irresponsible abuse of the creation of money and credit.

Spending cuts and tax cuts tied together with the pink ribbon of statesmanship have dazzled holders of the minidollar into believing that the American currency has finally left the primrose path of debasement.

Unfortunately, nothing could be further from the truth. The titanic abuses in the creation of money and credit since 1940 have irreversibly undermined the financial structure that underpins the minidollar. So now, even if a balanced budget were to be realized, it could in no way reduce the greenback's loss of purchasing power.

After four decades of maltreatment of the nation's money, the vested interests of inflation have become too powerful and too influential. They have absolutely no desire to see their financial empires and attendant political privileges dismantled by such an obscene thing as a monetary unit, stable in purchasing power as well as in foreign exchange value.

69. Government bonds are certificates of guaranteed confiscation

I don't have a great deal of respect for the intelligence of the people who occupy positions of power in the US Treasury. I mean I can't ask that the bond salesman who is Secretary of the Treasury be familiar with the writings of David Ricardo.

Nevertheless, a few people with knowledge in the Treasury know that I speak the truth when I say the pyramid has to fall. So it becomes necessary to take steps to divert attention from the decay of the dollar.

The Federal Reserve System's financing of continued deficits through the purchase of government bonds is a problem that aggravates me very much. I was shocked when I finally realized that total public and private debt in this country now amounts to approximately 12 trillion minidollars. There is no chance of it being repaid with the same purchasing power originally invested by the buyers of these debt obligations. They are not bonds. They are certificates of guaranteed confiscation.

70. If the riots in South Africa continue, I am afraid the gold production will severely decline

I have been to South Africa numerous times, and when I came back the last time, I advised all my friends and clients to get out of all South African printed paper. And if the problems of riots and strikes in South Africa continue — and I talk with my sources there at least once a week — then I am afraid the gold production will severely decline. In fact, I believe South Africa will stop shipping gold altogether.

71. The currency problem, as I see it, is a rape of the law

The most serious problem we face today is the debasement of our currency by the government.

The first car I bought shortly after World War II cost me some $950, even with all the extras. The average car is now $10,000. I visualize that soon it will sell for $20,000.

My friends, I am a pupil of G. F. Knapp, a distinguished professor of currency theory who taught at the University of Strasbourg. Professor Knapp said that currency is a creation of the institution of law. And today, the currency problem, as I see it, is a rape of the law.

72. My tailor knows more about currencies than all the idiots at the US Treasury put together

Let me tell you a little story.

My first suit in this country was custom made for $75 by a Britisher who had just immigrated here to profit from the prosperity. At the time, I said to him, ''Mr. Stewart, one day you will charge me $500 for a suit, and I will pay you.''

Years later, in 1979, he charged me $900 for the very same suit.

Saturday morning, I went to see Mr. Stewart about a coat. He criticized the suit I was wearing and said, ''Dr. Pick, a man like you should have a new suit.'' I said, ''OK, what do you want to charge me?'' He held up two fingers.

''You don't mean two thousand dollars?''

He said, ''NO — two krugerrands!''

The craving for gold is increasing every day all over the world. My tailor, who is 81 years old, knows more about currencies than all the gang of idiots at the US Treasury put together. It is the destiny of paper currencies to lose their purchasing power, and I have seen nothing in the past year to change that fact. We are now living in the 45th year of the legal expropriation of every owner of paper dollars in which so-called ''securities'' are denominated, and I guarantee you, there will be more expropriation in the months to come.

73. By debasing the currency, we have endangered the economic existence of the United States

The destiny of the currency is the destiny of the nation. The two are the same. You cannot run away from it. Yet the government does not want to admit what it has done to the currency.

You see, the mistakes of the administration are constant. You note that I said mistakes; I don't say willful mistakes. We had the most disgusting adventure in Indochina. We sank about $600 billion into the war over there and lost. In doing so, we ruined the dollar, and yet we had daily victory bulletins. By debasing the currency, we have endangered the economic existence of the United States.

I am a poor man's child and had to work my way through various universities to gather the knowledge that provides my employment.

But for political expediency, the Truman Full Employment Act of 1946 has taught us that everyone has a right to a job.

The inflation we practice in order to avoid unemployment is based on a misunderstanding of the theories of John Maynard Keynes who said that during contracting business cycles, you have to go into deficit spending.

But Keynes also said that the <u>moment</u> the cycle expands again you have to repay the deficit. This part of his theory had to be discarded for political reasons, and we have had nothing but continuing inflation for decades.

74. Corporate profits, on a per capita basis, are only about a quarter of what they were in 1950

I want to discuss the economic superstitution that ''a little bit of inflation'' is a reasonable price to pay for an expanding economy.

The ''little bit of inflation'' is real enough. But the ''expanding economy'' is a sleight-of-hand illusion.

Pre-tax corporate profits figured in constant dollars peaked in the early 1950s. Since then they have declined steadily. Today, on a per capita basis, they stand about half of what they were in 1940, and only about one-quarter of 1950.

Today, even the newspapers admit that corporate profits are down. But that's a lie, an illusion. They are non-existent. The corporations today have no profits. The American corporation is living off its capital.

What I teach is heresy. If the true condition of the American corporation was known, no one would buy corporate stocks or bonds. That is why the newspapers will not print what I say.

The use of nominal minidollars in corporate balance sheets obscures the decline of the corporation. Deflate the capital base of any large corporation or bank today and recompute it in constant dollars. You will see they have no profits when you take into account currency debasement. You will see that their ''dividends'' are not dividends paid out of profits — they are return of capital.

75. I itch in my hand to buy gold now

I've been bullish on gold since it was $35 an ounce. The people who I put into gold bought at mostly below $100, and I still think they should sit on it because no government has ever mastered gold.

I don't have any long-term outlook for gold prices. I make my plans each day. But I believe we are temporarily at the end of the decline — 10% more or less. From here on, there are many questions for which we do not have the answers. Will Mr. Volcker resign? Will Mr. Reagan survive his term as President? Rumors are multiple. But in general, nothing much will be changed.

Let me say, I itch in my hand to buy gold at the present time.

76. The biggest single industry in the United States is corruption

For the past 40 years, I have been doing research on corruption. My first big assignment was to find out the volume of off-track betting in New York State because a friend of mine wanted to establish off-track betting officially.

During this 40 years, I have learned that the biggest single industry in the United States is corruption. It's about five times the volume of the automobile industry — tax exempt.

No one in office ever mentions the word corruption in any of their speeches. The more things change, the more they stay the same.

77. I do not give any country the right to limit the transferability of my currency

Currency controls never work. They automatically create a black market.

Do you know how old currency regulations are? Let me tell you where currency regulations came from and why. In 1909, Professor Knapp, whom I studied under long ago, was approached by the Imperial General Staff of the German army. They asked him to give them one of his best pupils to draft legislation for currency controls because they were afraid they would be involved in the Balkan wars as allies of Austria.

As a result, on August 1, 1914, orders went out in Austria and Germany for citizens to surrender their gold and hard currencies to the government. We never had currency regulations before. Made in Germany. And now, the IMF publishes a huge book, one inch thick, full of all the world's currency regulations.

I ask you, do I have to accept currency controls if they ruin me? I say no. I do not give any country the right to limit the transferability of my currency or to prohibit me from owning better than the national currency.

I am not alone. That is why currency controls never work.

78. The top 3 or 4 banks will be officially nationalized

Back in 1958, I got aggravated about the dollar and didn't come into the office for five days, and on Monday I came in with a manuscript titled ''The United States Dollar — a Requiem.'' The boys read it, and said it was good but I couldn't use that title.

I asked why not, and they said that the dollar's not yet dead. Wonderful. I revised the title to ''Requiem for the Dead Half.'' Now it's just about time for my original title.

Whether the government likes it or not, it will have to come down to earth and admit that through inflation, they expropriated the money people put into savings banks, pensions, and government bonds. It will be necessary for them to say they were foolish and that now is the time to start all over again with a decent currency.

The government will wipe out all the debts. But since every debt is someone else's asset, many innocent people will be ruined.

Take life insurance for example. Insurance is a good business for the insurance companies...but not for you or me. Because the insurance company owes minidollars. And they pay off in depreciated dollars. When the currency exchange is announced, it will just confirm the fact that the currency they owe you is worthless.

I believe the top three or four banks will be

officially nationalized at that time. They're practically nationalized now. They can't even sneeze without government approval. No profits flow from them in terms of constant dollars. Nothing serious will happen to the rest of the banks.

79. The Eurodollar has to be compared to the second use of toilet paper

I am afraid I will have to be very naughty in what I tell you about the Eurodollar. We created the Eurodollar because we were unable to cope with the problem of post-war financing.

The Eurodollar is nothing but the accumulation of the deficits of the balance of payments of the United States. And, if the deficit serves the refinancing, the Eurodollar has to be compared to the second use of the already used toilet paper.

I'm not very welcome with these statements, but I have no favor to ask from anybody and at the age of 87 years, I will not be jailed for high treason.

Few people know anything about the Eurodollar. But understanding the Eurodollar is essential to understanding our total currency cancer infection.

80. We will have a new "hard" dollar on a gold standard

There has been some talk of restoring the gold standard in this country. However, no politician wants to carry the load of responsibility for maintaining the monetary discipline imposed by the yellow metal.

Nevertheless, the return of gold to some sort of monetary function is necessary to stabilize the currency in terms of purchasing power as well as foreign exchange.

Fundamental and necessary to such a stabilization of the dollar would be a drastic currency reform that would sweep under the carpet all the debased debts — both governmental and private.

If such a step were taken, a new ''hard'' dollar with some link to bullion would be created. This act alone would restore much of the power America has lost since 1945.

In addition, a temporary period of currency stability would begin. This would be welcomed by the public as well as by the financial community, and would also make it possible to balance the budget expenditures against revenues.

The international repercussions of such a currency reform would again restore some of America's lost prestige. And it would be worth all the hardships such a monetary reconstruction always brings.

81. Why corporations should use constant dollar balance sheets

The idea of a balance sheet in constant dollars is nothing new. It arises as a necessity whenever a currency loses its value. The man who calls himself an ''investor,'' the completely immoral so-called ''securities analyst,'' the administrator and operator of mutual or pension funds, the corporate president — all bear responsibility for cheating the public and, what is more tragic, for cheating themselves. They do not want to acknowledge that between 1940 and 1984, by <u>official</u> government figures, the US dollar has lost 86% of its domestic purchasing power.

Austria, Germany, and Hungary in 1920, France in 1922, and Czechoslovakia and Italy in 1923 adopted constant dollar balance sheet methods. Large companies had to have some means of analyzing the real value of profits. All of these countries created excellent laws that filled volumes, whose purpose was not only the adoption of figures to new currencies, but also involved taxes, rents, debts, and liabilities. The currency adjustments worked smoothly and created few complications.

I have been trying for years to get governments, corporations, and individuals to revise their accounting methods in terms of constant dollars. I have learned only that I will never be recognized in the balance sheet field.

82. How currencies die

As currencies become more and more devoid of substance, they perpetuate their existence through their multiples. The <u>milreis</u> replaced 1,000 <u>reis</u>, and the <u>bilpengoe</u> tried to substitute for a billion <u>pengoe</u>. The <u>conto</u> was worth 1,000 <u>escudos</u>. The Greek <u>talent</u> was equal to 6,000 <u>drachmae</u> or 36,000 <u>obols</u>. In Java, the <u>bahar</u> was good for 100 million <u>candareens</u>. In India, the <u>nil</u> replaced one hundred billion <u>rupees</u>...

Some coins were flattened to the point where they were as thin as a sheet of paper, or actually chopped up into strips, or cut into bits of all sizes and shapes. Some were punctured and the holes then plugged with inferior metal.

These stratagems did not and will not save currencies, which are all doomed by the passage of time.

83. The pious pronouncements to hold the money supply in check will not be kept

The fellows in the central bank make pious pronouncements about fighting inflation and holding the money supply in check. But they panic immediately when they see signs of distress-borrowing in the banking system, as debtors — many of whom are corporations having interest payments larger than their pre-tax profits — try to keep their enterprises from going under.

Although the Federal Reserve system makes a lot of noise about controlling the money supply and reaching monetary targets, it is at times difficult to understand just what exactly they are controlling. Be that as it may, they will in time revert to form and resume the process of what is coyly referred to as ''reliquifying the economy.''

This will lay the groundwork for another cycle of currency destruction, which could assume unprecedented dimensions. Though ''to deflate or not to deflate'' may be the question, the only answer to America's growing financial and economic malaise is to debase.

84. Bond salesmen's propaganda that "a dollar is a dollar" should be rewritten to say "a dollar is 3¢"

Since most ordinary people, bankers, and company presidents have never studied currency theory, they swallow it hook, line, and sinker when the bond salesmen tell them ''a dollar is a dollar.'' That piece of propaganda should be rewritten to say ''a dollar is 3¢.'' The nominal dollar is officially worth no more than 14¢ of its 1940 value, unofficially only 3¢.

If computed in 1940 constant dollars, not more than $1,380 exists of the US $46,000 per capita gross public and private debt. More than $44,628 has been destroyed by inflation. But sadly, the owners of this debt do not want to hear about it. They do not wish to know that bonds are issued by governments with the sole purpose of debasement.

To my knowledge, no government in history has paid its debts in currency equal to the purchasing power of the currency lent to them. The people always lose their money on bonds.

It angers me. Bond salesmen should be thrown into the East River.

85. The US government should build a monument to Johannes Gutenberg

I have three unofficial deflators: the General Middle-Class Index; the Upper Middle-Class Index; and, for the politicans, labor czars, pimps, and TV idols, I have my Rich-Corrupt-Underworld Index. These indices show the dollar is worth maybe 3¢ of its 1940 value. If it's worth a nickel, call me a liar.

The $1,000 US savings bond is now worth only $30 in unofficial constant 1940 dollars. This is how the government expropriates the wealth of the people who trust it.

However, Washington has not tried to compensate the patriotic buyers of bonds for their loss of purchasing power. They didn't even get an income-tax reduction for the debasement of their bonds.

The US government should build a monument to Johannes Gutenberg — the man who invented the printing press 500 years ago. They should put it up in front of the Treasury Building which should be renamed after him.

People think of Gutenberg as the printer of great Bibles — which indeed he was. But in doing so, he also created the machinery for expropriation.

86. The triumph of gold

Throughout history, from the days of the Pharoahs down to Mr. Baker, gold has always been man's best protection against dishonest government money management.

Gold is money. Gold has always been victorious over futile government attempts to dethrone it. The history of man's attempts to conquer gold is fascinating reading. But in the end, gold always wins, because gold has no master.

May I take you back to the Vietnam war? Each day for over 4 years, we had a victory bulletin — until someone added them all up and discovered that many more Vietnamese people had allegedly been killed than ever existed. Nonetheless, we had victory bulletins. Now we are into a war on gold.

Attempts to dethrone gold by the US Treasury and its wholly-owned subsidiary, the International Monetary Fund, have accomplished nothing. The price rapidly regained any losses from the IMF and US gold sales. The US gold sales were tragic comedies featuring black humor, especially when you realize that the US has — it says — less than 8,600 tons, while the Russians have 9,000 and more likely 9,300 tons of gold.

The US war on gold was met with defeat at every turn. You know the record as well as I do, and there is little need to review all the events that have transpired since the collapse of the London Gold Pool in March 1968 down to the

current pathetic whimpering attempts to
persuade governments _anywhere_ to sell gold — no
matter how small or how insignificant to the
world market.

Now, let me declare openly and on the record
that the US Treasury has lost the war against
gold, and it doesn't want to admit it.